THE SCIENCE OF ILLUSIONS

JACQUES NINIO

This translation was published with the help of the French Ministry of Culture.

English translation first published 2001 by Cornell University Press
First printing, Cornell Paperbacks, 2001

Printed in the United States of America

Library of Congress Cataloging-in-Publication Data

Ninio, Jacques.
[Science des Illusions]
The science of illusions / Jacques Ninio ; translated by Franklin Philip.
p. cm.
Includes bibliographical references and index.
ISBN 0-8014-3770-9
1. Hallucinations and illusions. I. Title.
BF491 .N56 2001
153.7'4—dc21 00-011899

Cornell University Press strives to use environmentally responsible suppliers and materials to the fullest extent possible in the publishing of its books. Such materials include vegetable-based, low-VOC inks and acid-free papers that are recycled, totally chlorine-free, or partly composed of nonwood fibers. Books that bear the logo of the FSC (Forest Stewardship Council) use paper taken from forests that have been inspected and certified as meeting the highest standards for environmental and social responsibility. For further information, visit our website at www.cornellpress.cornell.edu.

Paperback printing 10 9 8 7 6 5 4 3 2 1

CONTENTS

THE SCIENCE OF ILLUSIONS

1 INVENTORY

The illusion of always being right.

The sound of the ocean in seashells.

The green ray: when the sun sets on the sea, after a beautiful day, when the horizon is sharply defined and the sky very clear, "at the instant immediately following the disappearance of the upper edge of the sun's disc we sometimes see a ray absolutely green, of great beauty, that follows the final red rays projected on the waters and in the atmosphere." (Trève)

On the sea, the waves appear to move, propelled by the wind, whereas the water merely rises and falls.

"When we stand motionless on the seashore at the spot where the waves dissipate, it sometimes seems that we slip with the ground toward the sea when a wave ebbs." (Bourdon)

"But when the sun was at last on the point of disappearing on the horizon, when its rays, much softened by the evening's haze, covered with the most beautiful crimson the surrounding world, the shadow changed color and appeared a green that, by its limpidity, could be compared to the green of the sea, and by its beauty to that of the emerald." (Goethe, describing colored shadows)

The brilliant spray of fireworks, after the explosion, goes out in all directions. But all the bursts seem to stream back toward us.

"The star is seen as larger than the point of a needle." (Armand de Gramont)

Traffic jams: the illusion of being in the slowest lane.

When two objects are moving at the same speed, the one farther away seems to be going more slowly. (Euclid)

Seated in a train that has stopped, one has the sense that it is starting up whereas it is only the train on the next track that is pulling away.

Upward variant: . . . one day, in winter, when there was no wind and a very heavy snowfall, my daughter, who was at the window, suddenly cried out that she was rising with the whole house toward the sky. (Ernst Mach)

When I have just stopped my car at a red light, I sometimes have a—very disturbing—sense of starting off in reverse, an apparent movement that persists despite my forceful braking.

The illusion of riding in the direction of the sun whereas the earth is rushing us headlong in a transverse direction.

Feeling tickled before being touched.

The sensation of having wet hands while doing the dishes, whereas one is running water over one's waterproof gloves.

When your lip is swollen by a pimple and you take a drink from a glass, the glass seems to have a warped rim.

The sensation of walking on unstable ground after a long bicycle ride.

Thinking on awakening that one's arms are in a certain position and finding them in quite another one.

The sensation of weightlessness that one has in free fall.

The illusion of the beheaded: floating outside one's body and seeing it as lying beneath.

"When you are strolling in the fog, a man encountered appears a giant because he is seen confusedly and as very far away; yet being close by, he projects a large image in our eyes." (Le Cat)

Arctic illusions: The monotonous surfaces of the Arctic create frequent problems in the perception of size and distance, especially in overcast weather. . . . A Swedish explorer had almost finished recording in his notebook the description of a steep mountain embanked by two curious symmetrical glaciers, the whole dominating a large island, when he discovered he was observing a walrus! (Barry Lopez)

Pathology: Dr. Lépine recently had in his office a woman thirty years old . . . who constantly *heard* a series of twenty-five words uniformly and regularly following each other without any apparent meaning. This woman had the distinct impression that these words were not pronounced aloud, and yet she heard them and, strangely enough, she heard them not in her ears but in her left cheek.

The people on the platform, seen from the train arriving at the station, seem small. Of two objects with equal weight, the smaller one seems to be heavier. Damp cold seems colder than dry cold. A coin placed on the skin of the forehead seems heavier when it is cold that when it is hot. (Weber's law) Twenty percent off the price of a second-hand pair of socks gives more pleasure than one percent off the price of a river of diamonds. (Fechner's law)

The fire-eater: seeing the fire coming out of his mouth while in fact he spews gasoline that catches fire outside. Hearing the song come from the singer's mouth whereas it is emitted by loudspeakers behind our ears. The illusion that the television announcer is looking us straight in the eye whereas he is reading the text being scrolled on a screen linked to the camera. The expressive portrait that follows us with its eyes.

Visions when you're half-asleep or hypnagogic dreams: "I was reading aloud *Voyage in Southern Russia* by Hommaire de Hell. I had just finished a paragraph when I instinctively closed my eyes. In one of these brief instants of somnolence I saw, hypnagogically but with lightning speed, the image of a man dressed in a brown robe and wearing a hood like a monk in the paintings by Zubaran. This image immediately reminded me that I had closed my eyes and left off reading: I abruptly reopened my eyelids and resumed my reading. The interruption was so brief that the person I was reading to did not notice it." (Alfred Maury)

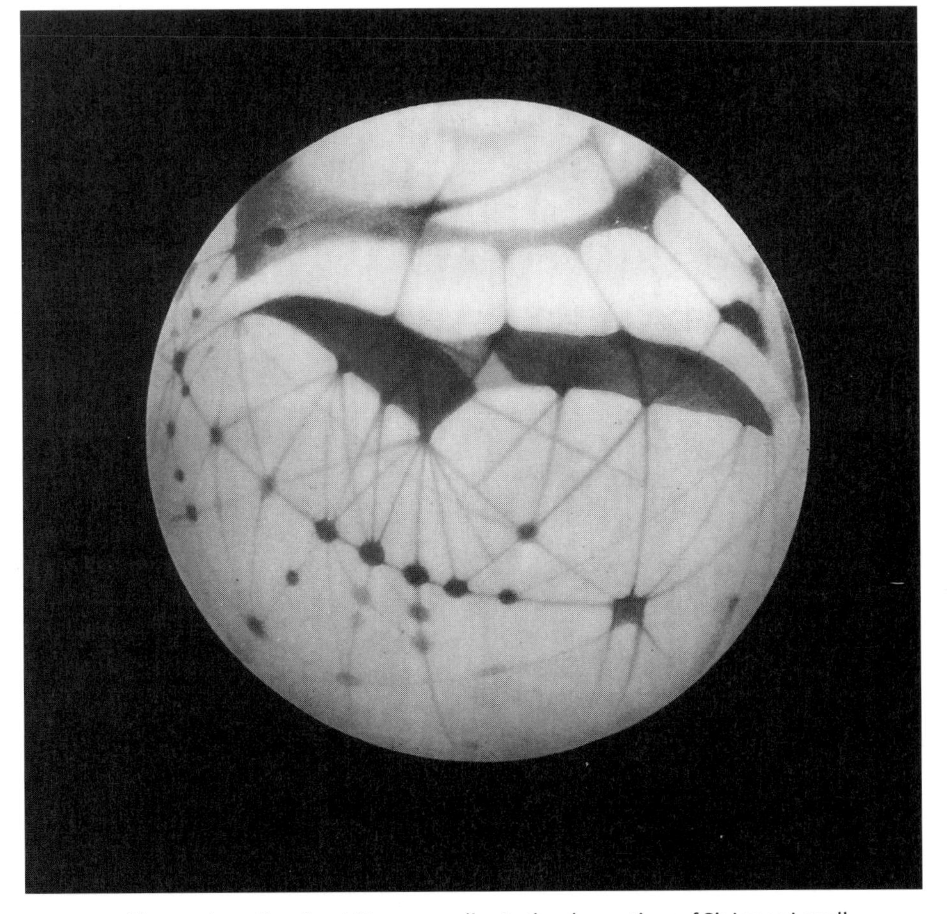

FIGURE 1-1. The canals on the planet Mars, according to the observations of Sir James Lowell. Their apparent geometric regularity is an illusion: "In these lines whose layout was taken for that of canals, we should see merely the surface features that happen to be laid out according to certain alignments. For the most part, because of inadequate optical power, they remain at the limit of perception and the eye, incapable of individually making out each of these little patches, sees only the confused combination of them resulting in the appearance of a continuous trail" (Lucien Rudaux, *Sur les autres mondes* [Paris: Larousse, 1937], p. 138).

The ticktock of a noisy watch: the time interval between the "tick" and the "tock" seems shorter than between that between the "tock" and the "tick."

A touch-tone telephone seems to start the call to the recipient before we have finished dialing his number. And we think we hear the ring at the other end of the line whereas the sound of the ring we hear is unrelated to the one heard by the person being called.

Don Quixote's windmills, Lady Macbeth's bloodstains, the emperor's new clothes, the man who mistook his wife for a hat, the invisible man, time travel.

The abominable snowman, flying saucers, the Loch Ness monster, the canals of the planet Mars. Seance tables, levitation. The philosopher's stone, the "sure thing," the foolproof strategy for winning at roulette.

Perpetual motion, the squaring of the circle. Sniffer planes, N-rays, the infallible method for losing weight, DNA beauty cream. The gene for human longevity, the intelligence gene. The cancer vaccine (cited on television), "clean" gene therapy. Physiognomy, a pseudoscience that claimed to gauge the character of an individual by his facial features. Dr. Bogomoletz's longevity serum, prefrontal lobotomy, a pseudo-treatment of mental illness causing real cerebral lesions (Nobel Prize in Medicine, 1949).

The child who tries to catch the moon.

The playground: from the other side of the wall, one hears only a piercing collective scream that rises and falls like a wind blowing through the branches of a tree.

Young people have no illusions about politics. (newspaper headline)

A mountain seen from a high point on the other side of a valley looks higher and steeper than when seen from below.

With age, we see inclines and stairs become increasingly steep. The garden where one walked as a child has shrunk.

A room emptied of its furniture looks smaller.

The illusion of being opaque and compact. Packed tightly as in a neutron star, all humankind would fit into two cubic centimeters.

To read a letter from you and to have the impression that you are speaking to me: when I am reading a letter from a person I know well, I hear him pronounce, with the exact intonation of his voice, the words that I am reading, and that he may never have said in my presence.

The illusion of reciprocity: to think that I must please her because she attracts me, to think that what seems to me brilliant must, at a minimum, interest her; to think she is smiling at me because I please her whereas she is smiling because she knows that she has me on a string.

The illusion that you give of listening to me and understanding me while your mind is elsewhere and the one that I give of being deeply asleep while I am solving geometry problems.

The illusion of your being the only one who can make you happy.

A BRIEF HISTORY OF ILLUSIONS

An illusion arises when one becomes aware of a discrepancy: on the basis of information supplied by the sense organs, we believe things are a certain way, but we know, through culture and reasoning, that they are not that way. In the face of such discrepancies we commonly accuse our senses of misleading us. However, they are not always wrong. The ideas that man forms about the laws of nature evolve in the course of history; what has been held as true becomes false and, as a result, what was thought an illusion becomes a legitimate representation of reality. Here are some phenomena cited as illusory by Lucretius, from the time of the Romans:

> Images in the mirror, echoes in the mountains.
> In sunny weather, our shadow follows us.
> Seen from afar, the square towers of cities look round.
> The sun and the stars seem motionless.
> On a moving ship, it is the countryside that seems to move. When a child spins around and then stops, the scene seems to turn 'round him.
> At night, in a cloudy sky, it is the moon that seems to move toward the clouds.
> Dreams and hallucinations.

Lucretius also described as illusory the appearance of a monument as seen in perspective:

Look at a portico supported by parallel columns all of the same height; if it is long and we look at it from one end to the other, it gradually narrows and takes the form of a cone lying on its side; the roof meets up with the floor, the right side with the left, until the eye confounds everything in the obscure point of the cone.

Nowadays it is a child's pleasant privilege to be fooled by the shadows that precede or follow him, that become longer or shrink as he approaches a streetlight, and that split in two or go off at right angles when he turns. I remember having a strange feeling when I was five, during a car trip, riding over flat land on a straight road through the desert; that day I became acquainted with a new sensation: the sides of the road were first parallel, then met in the distance—but before the horizon?—and their meeting point constantly receded as we advanced.

Was the common man surprised by the changing aspects of a temple? The illusion of a temple in perspective existed only within the framework of an erroneous theory of vision, possibly reserved for intellectuals of the time. According to this theory, objects emit eidolas, impalpable copies or "simulacra," that travel through the air and then penetrate the eyes. To see was to absorb eidolas, which had no reason to be affected by the distortions of perspective.

Euclid was well acquainted with the laws of projection from which the laws of classical or "linear" perspective follow. But in his *Optics*, written around 300 B.C., he also thought it necessary to formulate laws about the manner of seeing, such as:

(1) For a horizontal surface located above eye level, the parts farthest away look lower (the floors rise and the ceilings descend).

(2) An arc of a circle placed on the same plane as the eyes ("seen on the edge") looks rectilinear.

(3) As the eye approaches a sphere, the visible part will be smaller, but will look larger.

(4) If a sphere has a diameter less than the distance between the eyes, the viewer will see more of it than a hemisphere.

The first three observations are consistent with the laws of perspective, but the fourth one is more subtle. It results from the fact that the two eyes do not see exactly the same portion of the sphere. The central part seen in common and the two side parts, each seen with a single eye, are brought together seamlessly in the internal image we have of the sphere.

Euclid thought that the eye saw by "casting looks." His argument was that one can search for a long time for a pin that has fallen on the floor, without seeing it. He thus imagined a spray of rays leaving the eye, very close together at the start, but spreading apart as they traveled toward remote objects. To see the pin, one of the rays had to strike it.

Euclid was tripped up by one problem of which the importance is hard to appreciate today. The propagation of light in a straight line was easily observable in the case of light coming from a small light source. What was not at all understood was what could be the image of a large object that was not luminous in itself. Missing was a rather subtle concept that was to be introduced much later, around 1040, by Ibn el-Haytham in his *Kitab el-Manazir,* the book of visions, which was both a treatise on optics and a treatise on visual perception, one very much ahead of its time.

Ibn el-Haytham's (al-Haytham) important idea was that each of the points of a lighted object becomes a source emitting light in all directions. But through a suitable optical system (a plane mirror, for example), all the rays coming from a point on the object converge at a precise point on the surface where the image is formed. This idea, which had not yet found its correct mathematical expression, made possible a qualitative understanding of what would be the image formed in the eye. Ibn el-Haytham made some mistakes, however, that were cleared up for good by the astronomer Johannes Kepler in 1604.

Eclipses were particularly important for refining astronomical models: exactly how long it takes one star to "cross" another provides valuable cues about their relations of size and distance. It was still necessary to determine exactly the moment at which, in appearance, the two stars began to touch each other's outside edges. This determination was to prove particularly arduous: on the one hand, the imperfect lenses of the instruments tended to enlarge the celestial objects, and on the other hand, the observer was prey to a visual illusion that makes objects look larger when they are luminous. It was by tackling this twofold difficulty that Kepler sought to understand the optics of the eye. He proposed that the image is formed on the retina and that it is inverted in relation to the external world. As to what happens beyond the retina, he denied responsibility: "Optical scientists do not commit their troops beyond this opaque wall that constitutes the first obstacle in the eye." Beyond that, what we would call a signal is formed that remains separate from the object seen, and it is through the impression of this physical signal that vision occurs.

With Kepler, Snell, and Descartes, the propagation of light rays was well understood and the effects of mirrors and lenses were explained or were on their way to being assumed so. The phenomenon of the stick that is dipped halfway in water and that looks broken is not an illusion. Descartes explained how the rainbow is formed.

The ancients had noted that the rainbow originated in areas of the sky darkened by distant rains and could be seen by an observer only with the sun behind him. Hence the correct idea that raindrops act as mirrors. For Seneca, the rainbow was a myriad of images of the sun, reflected in each raindrop, and the rainbow was very large because the cloud would act like a concave mirror.

A ray of sun hitting a drop is refracted, then possibly reflected off the posterior face, is once again refracted and reenters the air. Descartes's contribution was to explain why the rays returning to the observer come from directions that form a large arc in space. The rainbow is a physical phenomenon, but one relative to the point of observation. Two observers near fountains where rainbows are formed do not see them in the same place.

In 1666 Edme Mariotte discovered the blind spot. This is the area of the retina from which the optic nerve leaves. He suspected that this area cannot contribute to the formation of the image, and he proved it. (Figure 2-1) Thus, we have a hole in our visual field that is filled in without our realizing it.

Jacques Rohault devoted a large part of his 1671 *Treatise on Physics* to the five senses. He astutely marked out the dividing lines between physical and perceptual phenomena. Like others of his time, he was attentive to the geometrical aspect of vision and enumerated the indices that make it possible to

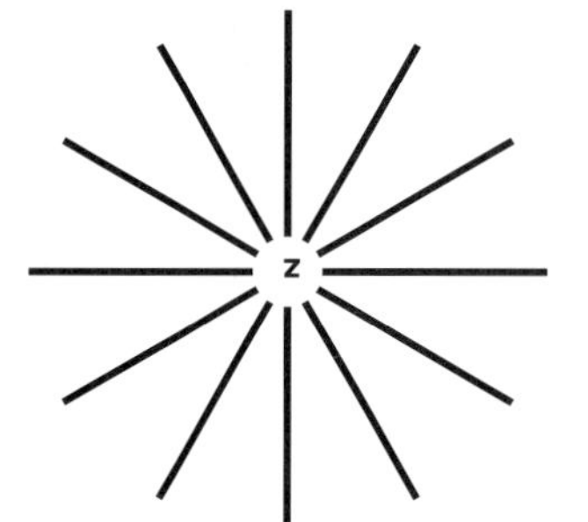

FIGURE 2-1. The blind spot, discovered in 1666 by Mariotte. Position your left eye some thirty centimeters above the black disc, and turn your eye toward the small X on the right. There is a position of the eye at which the black disc disappears. Similarly, with your left eye above the letter Z turn your eye to the right by the same amount; the Z disappears and the radiating lines look unbroken.

estimate the size and distance of objects and that sometimes lead to error. He taught us in passing that circulating at the time were "grooved images that represent various things, being looked at from different places":

> Thus one of these images being looked at head-on makes one see a Caesar, looked at from a certain side, makes a cat appear & looked from another side there is a skeleton: for as it is various parts of the image that cause these various appearances, so it is various parts of a pigeon that make us see various colors.

But Rohault also introduced new concerns—he was interested in the color changes of bird feathers, in the luminous streaks formed on polished metals, and especially in the sprays of light formed when we squint facing a

414 TRAITE' DE PHYSIQVE,
au haut de la retine, à l'endroit F L, cauſé
l'apparence des rayons d'embas CN; Si donc
on interpoſe le corps opaque OP, entre l'œil

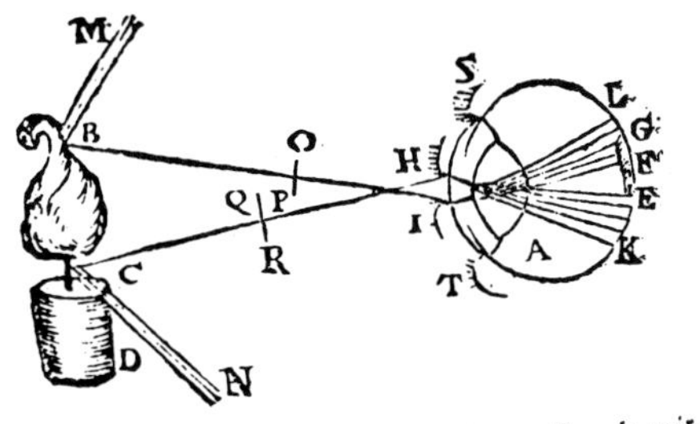

& le haut de la flamme, l'on ceſſera de voir
les rayons d'embas, & l'on continuera de
voir ceux d'enhaut, parce qu'ils ſont ſentis
par les rayons CH, qui partent du bas de la
flamme, & qui ne ſont point interceptez. Et
tout le changement qu'on experimentera en
ces rayons d'enhaut, ſera qu'on n'imaginera
plus ces rayons à l'endroit BM, mais ſeule-
ment ſur le corps opaque O P. Au reſte,
quand l'œil eſt ouvert à l'ordinaire, c'eſt à
dire, quand les paupieres n'avancent pas au
delà de S & de T, on ne doit point voir ces

PREMIERE PARTIE. 415
lumé, l'ô voit un cercle de feu aux endroits par où il paſſe? C'eſt parce que le tiſon fait impreſſion ſur des parties de la retine diſpo-ſées en rond, & que la viteſſe du mouvemēt fait que celle qui a été ébranlée la premiere côſ..e encore quelque peu de cette impreſ-ſion, quand le tiſon agit ſur la derniere.

De ce phénomene l'on peut conclure qu'encore que la viſion ſe faſſe en un in-ſtant, elle ne laiſſe pourtant pas de durer un petit eſpace de temps.

D'où vient qu'un boulet de canon, ou quel-que autre corps noir, paſſant extrêmement vîte au devant d'une muraille blanche, n'eſt point du tout apperçeu; C'eſt parce qu'un corps noir ne faiſant aucune impreſſion ſur les yeux, il interrompt alors ſi peu l'action des rayons de lumiere que la muraille re-flechit, que l'œil conſerve pendant un peu de temps l'ébranlement que les rayons ont fait immediatement auparavant.

D'où vient que quelques perſōnes ne ſçau-roient voir diſtinctement qu'à une certaine diſtãce, & voyent confuſément de plus loin, ou de plus prés? C'eſt parce qu'à force de s'être accouſtum és à regarder à cette di-ſtance, les muſcles qui devroient ſervir à changer la figure de l'œil ſont comme en-...dis, & incapables de faire leurs fon-

II. D'un tiſō agi en ron
III. Que la viſion dure quelque temps.
IV. Pourquoy on n'apperçoit pas certain corps qui ſe meuvē fort vîte.
V. Pourquoy certains perſon-

FIGURE 2-2. The light sprays according to Rohault. When your eye is turned toward a source of light and you close your eyes, you see rays of light form that start perpendicularly to the line of separation between the eyelids. Rohault noted that when one tries to cut the light flow by interposing one's hand or some opaque object (in Q), if the hand is toward the bottom, the low part of the spray persists but the upper part disappears. To explain this oddity, he assumed that the sprays are due to reflections in H and I on the edges of the eyelids. In fact, this involves a phenomenon of diffraction, which the physicists of the time could not have suspected.

candle—for which he proposed an incorrect physical explanation (Figure 2-2). Having noted a deficiency of color vision in his right eye, which was due to an accident, he speculated that some people could be born with such deficiencies. Concerning the physical understanding of colors, he lagged behind Newton. At the end of the seventeenth century Philippe de La Hire described—after Ibn el-Haytham—the phenomenon of color constancy: an object appears the same color in daylight and in candlelight; and in 1743 Buffon described colored shadows.

Meanwhile, the physics of light phenomena had made considerable progress. In 1669 Isaac Newton proved, with the experiment of two successive prisms, that white light is a mixture of lights of all colors. Physicists were then equipped to tackle complex optical phenomena: halos, the aurora borealis, parhelia (visions of multiple mock suns). From then on, detailed descriptions of meteorological phenomena proliferated in scholarly journals; in general, these were curious but objective effects simultaneously reported by several observers.

At the end of the eighteenth century the propagation of light in an optically variable medium was tackled. Gaspard Monge, who accompanied Napoleon Bonaparte on the expedition to Egypt, had a chance to observe some mirages. In general, mirages give an inverted view of the surface of the water that they reflect. Monge explained them as a stratification of the atmosphere into horizontal layers where the temperature increases on coming closer to the ground, which has the effect of curving the light rays and of making distant objects appear to be in other than their actual places.

There are also mirages that are not inverted, but right side up. Around 1820, a certain Monsieur Jurine described in the *Bulletin of the Philomatic Society* a curious phenomenon of a ghost ship observed on Lake Geneva, on which a small boat and its right-side-up mirage separated and sailed off in different directions:

> M. Soret saw appear above the water the image of two sails, which instead of following the course of the little boat, separated from it and took a different course, progressing . . . from east to west whereas the small boat was going from north to south.

The French physicist Jean-Baptiste Biot, very confident in the explanatory power of science, then undertook to demonstrate in the same *Bulletin* that this was exactly what he could have predicted. Taking account of the height of the sun above the horizon during the mirage, the topography of the lake, the layout of the nearby mountains, and the direction of the shadows, Biot estimated that the air had to be stratified into vertical slices that were colder near the mountain at the lake's edge.

At the start of the nineteenth century, new light processes were discovered: diffraction, birefringence, polarization. Physicists were able to explain all the bizarre behaviors of light. The iridescences, the streaks that appear in wiping the lenses of glasses, the luminous sprays formed when one squints facing a bright light no longer posed a problem. Then began the great era of experimental psychology. Many experiments were done using revolving disks. The interest was in stroboscopic effects, rapid successions of images briefly seen at regular intervals, and this led Joseph Plateau to invent a "stroboscopic disc" device for seeing drawings come to life. Photography, invented around 1830, helped in distinguishing the subjective from the objective. To the list of the illusions previously observed in nature were added dozens of illusions devised by scientists. Hermann von Helmholtz described a great many new phenomena, and between 1856 and 1866 he published his monumental treatise on vision, the *Handbook of Physiological Optics*, which still serves as a reference. In 1865, Ernst Mach proposed, in order to explain the illusion of the bands that bear his name, a model of interaction between neurons, a model that was to prove essentially correct. We had arrived at contemporary science.

The builders of the Greek temples were familiar with certain geometrical illusions. They tried to make up for these illusions by giving exactly what was needed by way of roundness and slants to their columns so that they appeared straight, and by adjusting their spacing so that they would look equidistant. In the sixteenth century Montaigne observed that a pattern of chevrons, familiar in heraldry and commonly used on rings, produces the illusion of widening on one side and narrowing on the other, whereas it has a constant width (Figure 2-3). In 1611 Rubens painted his *Descent from the Cross*. In it he deliberately shifted the two halves of a ladder interrupted by human figures in order to give the appearance of continuity. Much later, in the eighteenth century, landscape architects advocated planting trees to form hyperbolas on each side of the alleys in order to create an appearance of perfect alignment.

Around 1800 at the Ecole Polytechnique Monge taught descriptive geometry, an engineering discipline for studying forms in space, given their projections on two perpendicular planes. In 1836 Sir Charles Wheatstone invented the stereoscope or rather, one should say, stereoscopic imagery. Although the usual optical instruments (glasses, microscopes) are made to be trained on objects that already exist, the stereoscope is useful only in relation to images created for use with it, the stereoscopic pairs. These pairs represent two views of the same object as it would be seen by each of the two eyes. Through the stereoscope, each eye receives the image intended for it, and the sensation of three-dimensionality arises when the brain completes its comparative work using the minimal difference of perspective between the two images.

The human brain then appears like a gifted geometer, and interest focuses on the small mistakes it might make in the assessment of forms. In 1851 Fick published the first article explicitly presenting a geometrical illu-

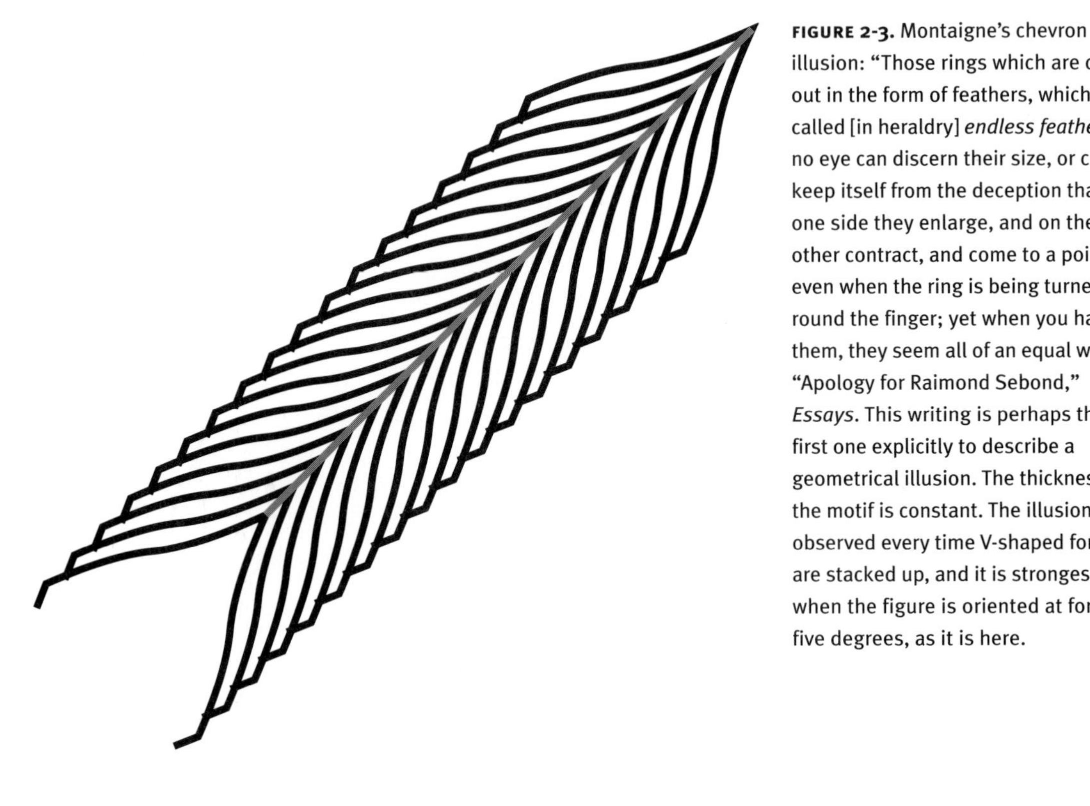

FIGURE 2-3. Montaigne's chevron illusion: "Those rings which are cut out in the form of feathers, which are called [in heraldry] *endless feathers*, no eye can discern their size, or can keep itself from the deception that on one side they enlarge, and on the other contract, and come to a point, even when the ring is being turned round the finger; yet when you handle them, they seem all of an equal width." "Apology for Raimond Sebond," *Essays*. This writing is perhaps the first one explicitly to describe a geometrical illusion. The thickness of the motif is constant. The illusion is observed every time V-shaped forms are stacked up, and it is strongest when the figure is oriented at forty-five degrees, as it is here.

sion where the vertical dimensions are overestimated in relation to the horizontal ones, followed in 1860 by Zöllner, who discovered an illusion similar to Montaigne's chevrons. An avalanche of discoveries followed them. Many geometrical illusions demonstrable with extremely simple figures were found, and they constitute to this day the main corpus over which debates are held about the reliability of the sense of sight.

The history of illusions of touch or audition is much less rich. For touch, Aristotle noted a doubling illusion (Figure 2-4), to which I confess I myself am not much susceptible: cross the two fingers, roll a marble in their intertwining, and you will have the illusion of rolling two marbles. It is also claimed that pinching one's nose between two crossed fingers produces the sensation of a double nose.

As regards audition, the ancients noted bizarre effects like echoes or the

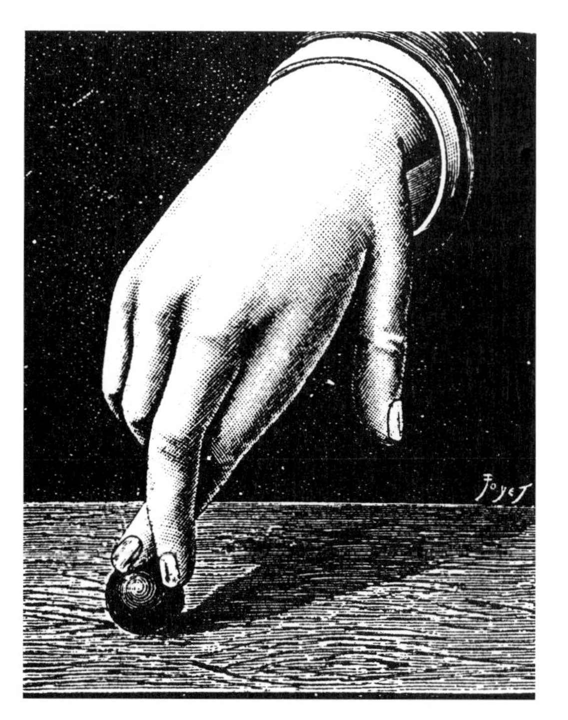

FIGURE 2-4. Aristotle's tactile illusion: when one rolls a ball between the index and the middle fingers, one has the illusion of touching two of them. Etching from *La Nature* 1 (1881): 384.

fact that we hear thunder after seeing lightning. They also noticed that the ear did not always locate a sound's origin in the right direction. The vibratory nature of sound was acknowledged, and people were acquainted with the phenomena of resonance: a string of a musical instrument is plucked and a string of another nearby instrument produces a sound. Jacques Ozanam, an author of scholarly books and of a popular treatise on recreational science in which he explained some "gamebag tricks," extrapolated somewhat when he mentioned around 1690 the synchronization of clocks:

> Here let us add a very curious thing: two pendulum clocks placed on the same shelf, will tend toward the same time, although their hands were not first set to the same time.

Mariotte was interested in the physics of gases and in the sound of the trumpet; he wrote a *Treatise on Percussion.* There he asserted that when two metal spheres collide, the spheres are compressed and lose some of their shape at the point of contact, that the metal acts like a spring and that the sound of the shock arises from that. Jacques Rohault, in his *Treatise on Physics,* devoted a chapter to sound and cited the illusion of the humming you hear when you tickle the inside of your ear. He also described an entertaining experiment performed by children (Figure 2-5): they suspend a metal clothespin on a string, twist the two ends of the string around two fingers, stick their fingers in their ears, and swing the upper part of their body to impart movement to the suspended clothespin. When the clothespin hits a hard object, people nearby hear only a faint sound, but the children hear a sound "similar to that of the largest bells in our churches." He drew this lesson from it:

> It is impossible to explain this in any other way than by saying that the moved clothespin disturbs the string, which transmits the impulse to the fingers, which next move the parts of the ear to which they are applied, stimulating in turn the nerves that make up the organ of hearing.

Around 1840, Doppler described the effect that bears his name: when an object comes closer, the sound produced seems higher-pitched; when the

FIGURE 2-5. The bell. Suspend in the middle of a string a clothespin or a metal spoon, place the ends of the string on your ears, and rock your head. If the suspended object happens to strike the edge of a table, "you will hear the sound of bells as loud as great cathedral bells" (de Savigny). The variant shown here is a toy that could be obtained in 1899 in Paris (according to "Scientific News" 11 [February 1899], supplement to *La Nature*).

object moves farther away, the sound seems lower-pitched. This phenomenon can be explained by physics. A short time later, acoustics became an established scientific discipline with, once again, the crucial contribution from Helmholtz, who analyzed musical sounds according to their fundamental frequencies and harmonics. Helmholtz noted some subtle effects in the perception of musical sounds, including the phenomenon of the missing fundamental: in certain instruments, notably the piano, the lowest notes do not contain the frequency that would correspond to the pure low sound whose name they bear. They contain the high harmonics of the fundamental sound, and the listener still hears the low-pitched sound because it corresponds to the interval of separation between the harmonics that are actually present.

Audition, which was studied very little by psychologists, was taken over by engineers, acousticians, musicologists, and certain linguists. Little was said about auditory illusions because, for one thing, the norms in audition were less clear; our conceptualization of the world and our vocabulary for describing the environment are based on vision. Auditory illusions began to be known—or constructed—in the twentieth century, in two waves, thanks to the contributions of an instrument—the tape recorder—and then of techniques for creating synthetic sounds.

Often, when a paradoxical effect is encountered in vision, we find the equivalent in audition. The two sensory modalities carry out information processing; they must extract an information-bearing signal on the basis of raw sensations; they must therefore increase the contrasts, and separate entities, to complete their identification, if necessary by relying on memory, and so forth. In this book I shall attempt as much as possible to present the different categories of illusions in their visual and auditory modalities. For vision, material is plentiful, and many effects can be demonstrated with fixed images. For hearing I shall refer to situations experienced in daily life and will not discuss at length those requiring laboratory setups, for they are not demonstrable on the printed page.

ONE ILLUSION HIDES ANOTHER

Illusions that strike us right away are rare in nature. In some cases, one has to force oneself to see things naively in order to become aware of an illusion. For example, facing a lake, a person knows that the surface of the water is horizontal and sees it as horizontal without asking himself any questions. How would he see it if he disregarded what he knows? In fact, it is fairly easy to see the surface of the water form an inclined plane. A bit of sea glimpsed through an opening between two mountains can look vertical, like blue water filling the bottom of a glass held in front of oneself. Bourdon drew attention to this effect for the sea, when he looked at it while tilting back his head: "Thus the sea observed from the top of some dunes then appears, if we are far from the shore, like a vertical wall."

In the course of a hike, we notice, as we progress, that the landscape changes proportions: in mountainous country, heading toward a hilltop, we find ourselves on a ridge in front of which extends a wide valley; the summit we were heading for then looks much farther away than we had thought when we had started out. From a high promontory, the mountains beyond the valley appear to have steep slopes. As we get close, the slopes of the nearby mountains appear gentler.

A scientist with an interest in these oddities, who noticed that they always happen in the same way, could devise an explanation for them and then put it to the test. For this, he could construct highly simplified situations in the hope of defining the phenomenon in all its purity—preferably in an image made up of three or four lines. Eventually, one would retain these images and forget why they were produced. The uninitiated person thus makes con-

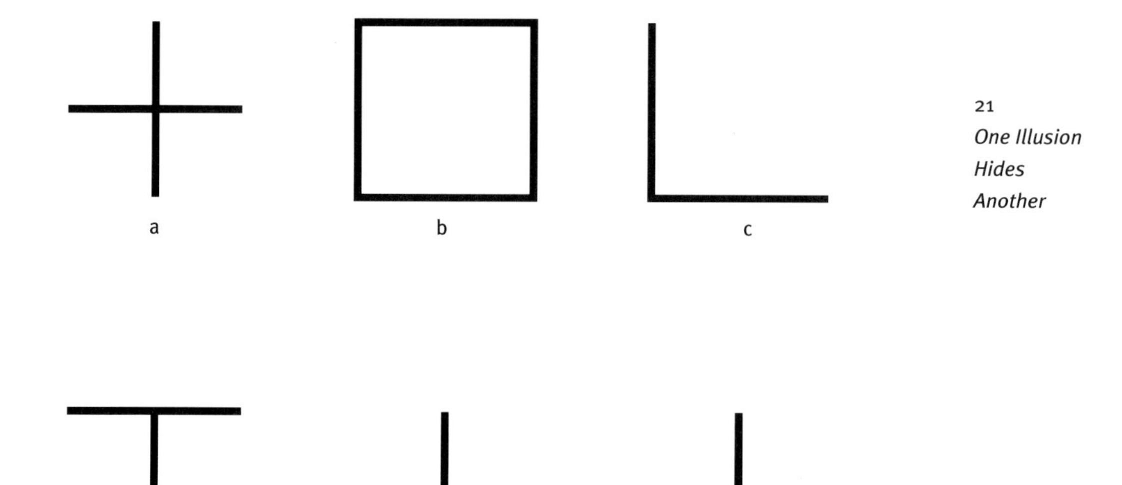

FIGURE 3-1. The horizontal/vertical illusion. The cross variant (a) was proposed by Fick in 1851. The vertical line looks a bit longer than the horizontal one, but the effect is not striking. If one is susceptible to it, one will find it in the square (b) and the letter L (c). In (d) the horizontal line of the T clearly looks shorter than the vertical line, but for another reason: any segment divided in two looks shortened. In (e) the horizontal line of the T is uninterrupted and the illusion is reduced or canceled out. In variant (f), proposed by Künnapas, the uninterrupted horizontal line is seen as longer than the vertical line.

tact with the illusions through images given as a set of oddities, nearly always the same ones. For the scientist, the illusion is a pedagogical example constructed to show an effect in a clear and measurable way.

Let us start with a common illusion, the "horizontal/vertical" illusion. The proposal is that a segment seems longer when vertical than when horizontal (Figure 3-1). This proposal is often illustrated with a letter T, which looks taller than it is wide, but whose two lines are exactly the same length. When we draw variants of it, we realize that the T example is a trick. An L formed with horizontal and vertical lines of the same length as those of the T looks less elongated. In fact, the illusion of the T is due for the most part to another effect, which is that a partitioned segment looks shorter than the whole segment. For example, the lines of a cross are seen as shorter than the parallel lines of the same length but separate (Figure 3-2).

In the nineteenth century the favorite illustration of this phenomenon was that of the top hat drawn to be as wide as it was high, but which looked nevertheless higher than it was wide (Figure 3-3a). This hat was suitable for another illusion that was also popular at the time: if you estimate the height of a hat a partner is wearing by making a mark on the wall near the floor (Figure 3-3b), you draw this mark systematically too high. Here the nature of the illusion is unclear. What is the source of the error? People underesti-

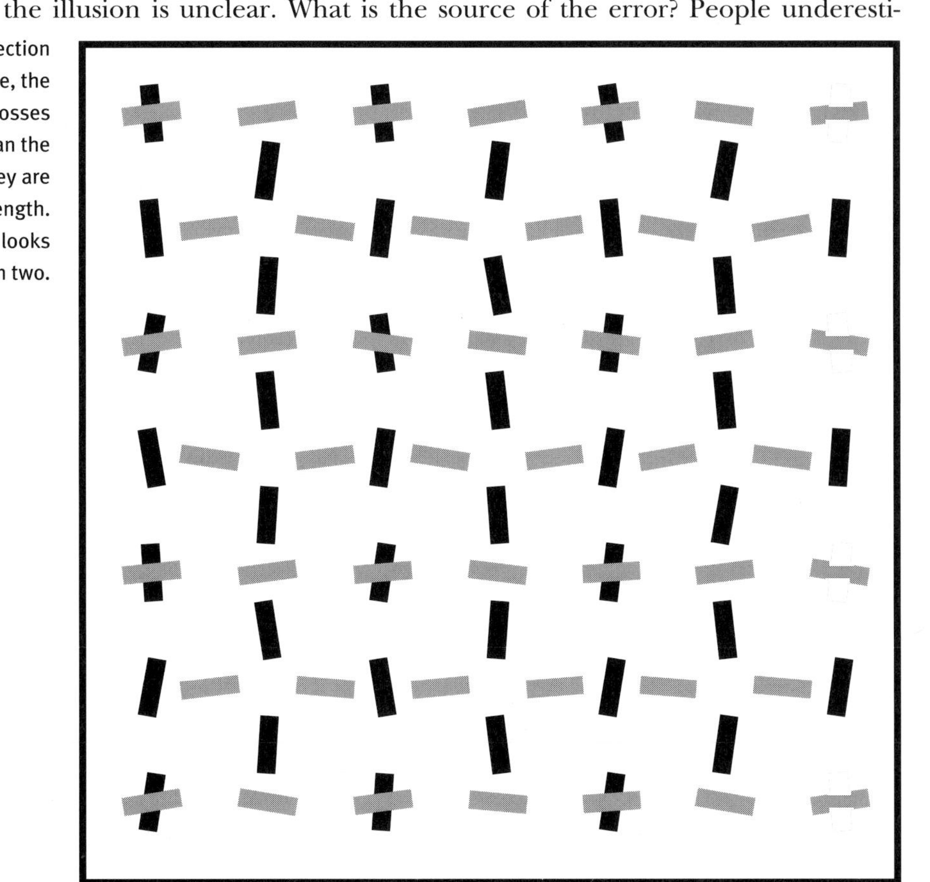

FIGURE 3-2. A bisection illusion. In this figure, the lines forming the crosses look shorter than the isolated lines, but they are all the same length. Generally, a segment looks shorter when split in two.

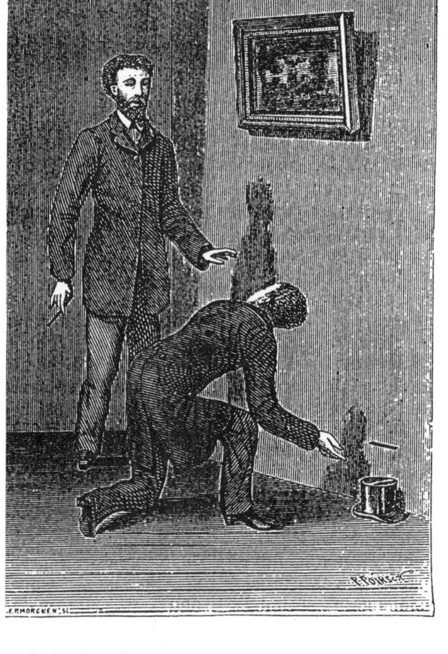

FIGURE 3-3. The height of a hat. This was a favorite example in the nineteenth century for showing the horizontal/vertical illusion. The top hat looks higher than it is wide, but its width, measured from A to B is greater than its height, measured from C to D. The engraving, taken from an unidentified magazine of novelties, appeared in *La Nature* in 1890. The illustration on the right appeared in *Science et Nature* in 1885, in an article entitled "Aberration of the Sense of Sight," which describes the engraving's experiment this way: "You ask a person to make a mark on the wall at the height, starting from the floor, of a top hat you have put on your head. . . . Nearly always, unless the person has witnessed the experiment previously, there is a difference in height that goes from a quarter more to the double and sometimes more." It is also reported in the article that if someone is shown a globe and asked to draw it to scale on the blackboard, nine times out of ten he or she will draw a circle that is 25 percent too large.

mate the height of the wall underneath the mark, in relation to the height of the hat. You see the hat head-on but you see the lower part of the wall by lowering your eyes. It seems that we generally underestimate the lower part of figures unless we are on our guard.

In the imagery of the twentieth century the example of the hat was replaced by that of two tables (Figure 3-4). One table looks long and thin, another one

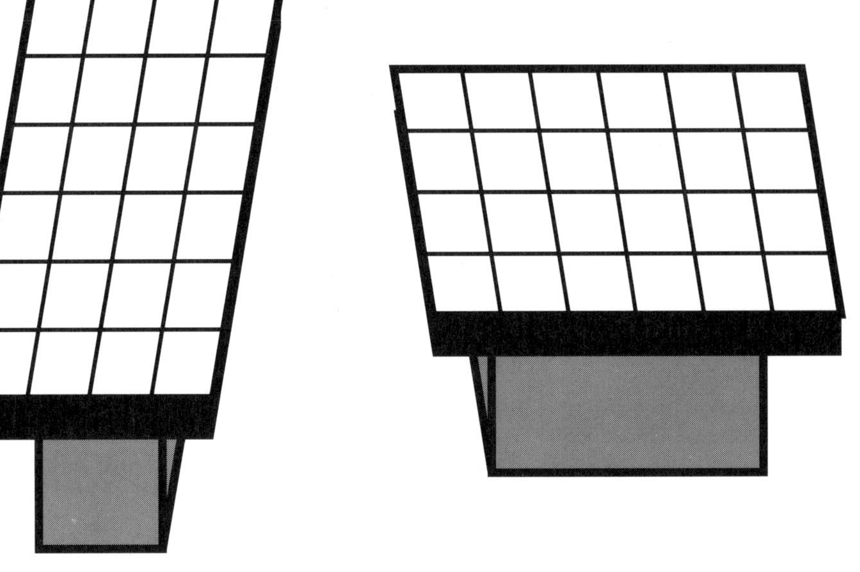

FIGURE 3-4. Shepard's tables. The tops of the two tables and their tilings have the same dimensions; they are exactly superposable. One could invoke the effects of perspective, but the illusion remains the same at all orientations, which can be confirmed by turning the page around. Curiously, after being turned over, the tables appear to have a perspective in the normal direction. Moreover, the illusion is also effective with simple parallelograms placed like the two tabletops.

short and squat. In the drawing, the tabletops have different orientations, but they are in other respects strictly equal. This is what would be predicted in the first analysis by the "horizontal/vertical" illusion, which leads us to see the objects as too long in the vertical direction. Another, more relevant explanation would be that the tables are seen in perspective. In its interpretation, then, the brain would tend to restore the correct form and thus deploy the figure in the direction of its depth, from which follows a widening of the transverse table and an elongation of the table in depth. But when we turn the figure around, the illusion remains, which does not square with either of the two explanations.

The vertical is first, in the absolute, the direction of gravity, which leads toward the center of the earth. In a drawing the verticals represent either these "terrestrial verticals" or the direction of depth: the farther away an ob-

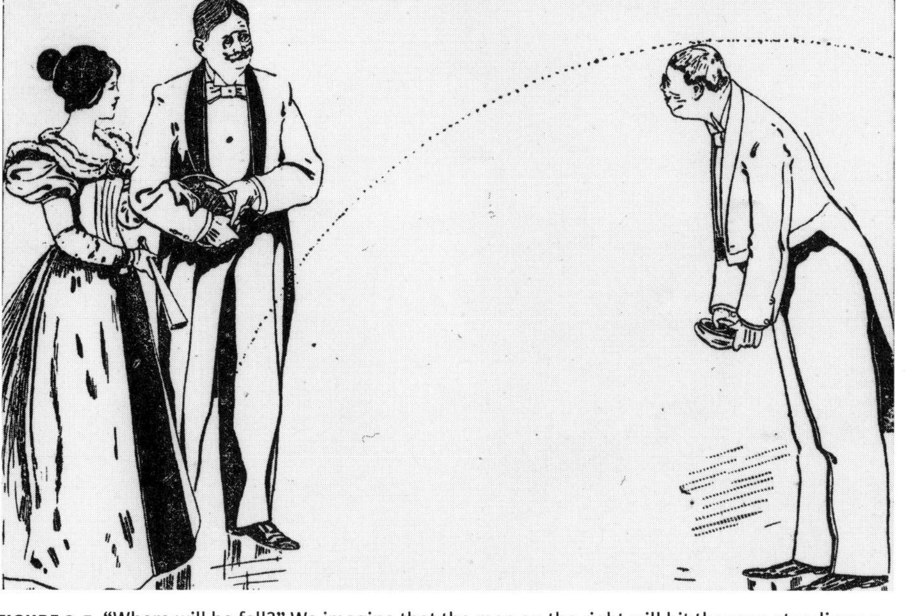

FIGURE 3-5. "Where will he fall?" We imagine that the man on the right will hit the man standing on the left at about the height of his knees, whereas in fact he has room to lie down full length. This engraving appeared in the book *Les Amusements de la Science* (Paris: De Savigny, 1905). Courtesy Bibliothèque Nationale de France, Paris.

ject is, the higher it is represented on the page. Thus there is confusion between "above/below" and "in front of/behind." In the illustration with the tables, the vertical concerns the above/below aspect for the base, and the in-front-of/behind aspect for the tabletops.

What becomes of them under natural conditions? Here I distinguish three directions: the width, in the direction of the line of the two eyes and that of the horizon; the height, which corresponds to the physical vertical of gravity; and the depth, which has to do with the observer's distance in the direction of his gaze. Concerning the classic width/height pair, the exaggeration of height is completely general. It is observable in real-life situations, as in photographs. A building facade that fits exactly into a square is seen as higher than it is wide. A man looks shorter lying down than standing (Figure 3-5). As

FIGURE 3-6. Hammersley's illusion. The open door looks wider than the closed door. At a close distance one would tend to attribute greater depth to surfaces than they really have. The width taken up by the open door in this photo is about half that taken up by the other door. The tendency to exaggerate the depth is reversed at medium and great distances.

regards the depth/width pair, at short distances we would overestimate depth in relation to the width (Figure 3-6).

The phenomenon is reversed at long distances. This is shown by a particularly strong illusion that to my knowledge has not yet been described; I call it the illusion of the arches (Figure 3-7). I happened to notice this illusion in a subway corridor. Its walls were painted with rather wide vertical stripes of different colors, forming the bases of arches that started on one side, passed overhead and came down on the other side. The nearest stripes appeared to be slanted at forty-five degrees, whereas the ones farthest away looked vertical. As I went forward on the moving sidewalk in the center of the corridor, the farthest stripes, which were first vertical, leaned over and took on a slant

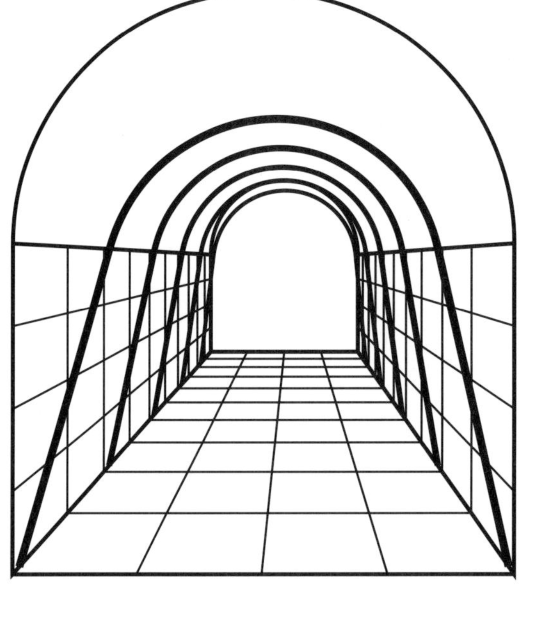

FIGURE 3-7. The arches illusion. In exact perspective, this drawing represents decorative arches along the inside of a tunnel. The arches are all parallel, but the farther away they are, the more vertical they look. We are well accustomed to the effect of the convergence of parallels in perspective, but much less so to the straightening of inclined planes in the distance.

of forty-five degrees to the rear. In actuality, all the stripes were in parallel inclined planes (see Figure 3-7).

The illusion was that the planes of the arches seemed to meet up, whereas they were actually parallel. But this illusion is consistent with the laws of perspective. Once we are alerted, we find it in some familiar images—for example, in the photos of stadiums or pyramids. One reaches the summit of the pyramids built by the Maya by means of stairs all in one block, which go straight up from the base to the top. Seen from the front, these stairs look vertiginously steep, and climbing them seems to demand a certain stamina. The inclination of the faces of the pyramid, which bear the steps, is better assessed from a side view, from which it is clear that the stairs are much less steep than they had appeared. The same effect is also observed in photos of stadiums or amphitheaters: the distant tiers facing us look steep, whereas the nearest ones have a gentle slope (Figure 3-8 and Figure 3-9). When we pay attention to them, the stairs in the form of a helix also produce the illusion: the steps we see head-on look closer together than those seen from the side.

FIGURE 3-8. Inclined planes. The distant tiers of the amphitheater, seen head-on, look steeper than the side rows. Notice the inversion of the relief when you turn the image upside-down.

Finally, it is for the same reason, presumably, that mountains seen from afar look steeper than those seen close-up.

Another familiar illusion is that of the images taken with a telephoto lens on a highway: distant oncoming vehicles look squashed from back to front. The effect of compression is due to the distance, the telephoto lens merely making the image come closer without distorting it.

At short distances we automatically correct for the effects of compression owing to perspective. For example, if I am looking at a tiled floor, the tiles at my feet, viewed perpendicularly, have the same form as the ones seen at an angle two or three meters away. Farther away, the perspective effects are no longer compensated for, and I see squeezed forms. Where does the transition occur? I suspect that the range over which the effects of compression are corrected varies with the size of the figure we are dealing with: possibly twenty times that size?

FIGURE 3-9. Stairs. In the left-hand picture, the stairway photographed from a distance looks much steeper than the one on the right, which was photographed close-up.

The assessment of slopes varies with age and physical state. According to D. R. Proffit, people estimate at twenty degrees the slopes of roads or land areas that are really only five degrees, and this error is purely visual: when they are walking, the movements of their feet are adjusted to the actual inclination. The rally cyclist completing his twentieth lap finds that the slope just before the finish line seems visually much harder than on the first lap.

Seen from the summit, the side of a hill clearly looks steeper than when seen from down below. The comparison is no doubt biased by the fact that, to judge the slope when one is at the top, one has to go to the extreme edge where the steep slope begins, and lower one's head, whereas to judge from down below, one can remain at a distance and gently lift one's gaze. Perhaps also our judgment may be influenced by the assessment of the physical difficulty involved in climbing up or going down the hill, climbing up being easier for slopes more than thirty degrees.

The illusion of the arches gives the impression that we are comfortable with the three principal directions of space, but that oblique elements, particularly if they are steeply slanted, give us trouble. The appearance of the geometric figures changes with their orientation. Even a figure as simple as a square produces some odd effects. Take a square whose sides are horizontal and vertical. Turn it forty-five degrees; it looks larger (Figure 3-10). Once it is turned, we are tempted to call it a rhombus even though it is still square, and it is often called a diamond.

Nearly all the geometrical illusions display important variations with orientation. When the illusion concerns the assessment of parallelism or alignment, it is almost always minimal for horizontal and vertical orientations, and maximal for orientations around forty-five degrees. (For example, see Figure 2-3.) It also seems that in judgments of size, we are less precise about the oblique directions. Hermann's grid (see Figure 6-5) produces qualitatively different illusory effects depending on whether its elements are oriented as squares or diamonds.

Except for these facts that I cannot explain, I have a sense that human vision is fairly well suited to coping with all the orientations. It handles the matter well whatever the posture of the body or the direction of the head. Its problem rather lies in its tendency to relate everything to the terrestrial vertical. I leave to the reader the task of estimating what happens to illusions when, instead of turning the figure forty-five degrees, the reader inclines his head.

Behind every illusion we find another one. The illusions form a huge collection in which we always end up finding a chain of family ties that connect the most disparate illusions. Having examined one of the simplest illusory figures (the letter T whose vertical line is overestimated relative to the horizontal line), we came across other geometrical illusions: the bisection illusion, the square/diamond illusion, and we have brought up the relevance of such illusions in a natural context, whether in the city or the country.

The visual illusions are the stakes of closely argued debates among specialists, and although there are differences of opinion among them, we cannot be satisfied with the first explanation to come along. The specialist has reliable means for eliminating certain hypotheses. For example, the vast ma-

FIGURE 3-10. Squares and diamonds. A square whose sides are horizontal and vertical looks smaller than a square of the same size rotated forty-five degrees. This can be confirmed for both the black shapes and the white shapes within them.

jority of geometrical illusions do not originate in the retina but at a later stage in visual interpretation, and the great majority of illusions have nothing to do with the eye movements performed in looking at them. The neophyte with no access to the technical arguments but who still wishes to test his personal explanation for an illusion is advised to construct many variants of it. With some imagination he stands a good chance of finding a decisive counterexample to his explanation. In the case of the square/diamond illusion, one could propose, as an obvious explanation, that the areas of the two figures are judged according to the horizontal and vertical directions. According to this idea, one would judge the size of a square by its sides, and the size of a diamond by its diagonals. However, in Figure 12-6 we notice that in the comparison of a large square and a small diamond, it is the square that is overestimated with respect to the diamond. This second square/diamond illusion is a sign, in any case, that even if the hypothesis was correct, it is far from explaining everything.

4 CLASSIFICATIONS

Philosophers interested in the workings of the mind would love to reduce all illusions to a single or a few general principles. William James's three-point schema provided a ragbag framework but one not without relevance. First point: brute perception, he says, does not exist; it is contaminated by memory. Do I have the watch in my pocket? Without putting my hand in my pocket, I feel it from the outside. The sliding of the material over the crystal convinces me that the watch is indeed present and reassures me. Such brief contact through the cloth of the pants must provide only partial information about the pocket's contents. All the same, the watch acquires an immediate presence through indirect contact with the hand, as if I had it in front of my eyes. In reading, one identifies a word on the basis of a few letters, and hence many printing errors pass unnoticed. Conversely, when we read a word letter by letter, paying attention to its typography, it becomes a strange object; its letters seem juxtaposed arbitrarily, and some doubt arises as to its meaning. A related effect, known as "verbal satiation," was described for audition by Edward Titchener in 1915. Repeat aloud a word—the first one that comes to mind: *house,* for example—over and over; soon the sound of the word becomes meaningless; you are puzzled and a little bemused on hearing it.

This effect was confirmed by other authors who added that, in this experiment, the meaning of the word rambles away.

The second point of James's schema is that in the case of doubt between two interpretations, A and B, perception never supplies a mixed interpretation. It always decides, proposing A, say, then switching to B. In a typical example, I have an appointment with someone. From a distance I see him in a crowd, coming toward me. I see his face clearly for a second or so. Then I realize my error, and the person coming toward me takes on a different face. People often make fun of mistakes made by the hard-of-hearing who, like Professor Tournesol in the adventures of the character Tintin from a popular comic strip by Hergé, instead of asking the speaker to repeat, interpret everything immediately but incorrectly. In 1936 B. F. Skinner prepared recordings of three to five vowels cyclically repeated many times. The persons listening to them first thought they heard three to five syllables of an indistinct conversation, then came to believe that they understood what the voice was saying to them; it concerned their private lives. The subjects were sure that their descriptions were accurate.

The signals provided by the sense organs are thus open to multiple interpretations: a great many stimuli could be the source of one and the same signal. Let there be a signal O and two sources, A and B, capable of producing O. Suppose that A is frequently encountered in everyday life. We get into the habit of assigning a signal such as O to source A without fuller examination. But under certain conditions, signal O can come from a source B that is different from A. From that arises an illusion that makes us perceive A instead of B. This is the third point of James's schema, which he refines by distinguishing two major type of errors: (1) object A is perceived mistakenly because it is "the most plausible, habitual, ordinary cause of O without being the real cause in the present case"; (2) object A is not the ordinary cause of O, but "it currently preoccupies consciousness, which is thus predetermined to let O suggest A."

In illusions of the first kind, James compares Aristotle's tactile illusion (see Figure 2-4) to that of the hollow mask (Figure 4-1) and includes illusions brought about by movement. In those of the second kind, the subject anticipates an event: the arrival of a loved one whom he is waiting for, or prey that he is hoping to kill, and, at the slightest sign, he thinks he sees the expected being. Suppose, says James, a sportsman hunting for a woodcock in a thicket sees a bird the color and size of a woodcock rise up and fly to the branches; he will instantly credit it with all the other features, though he doesn't have time to make sure; and he will soon be quite surprised and disappointed to find that the bird he brought down was only a thrush.

This classification by William James will appeal to the philosopher because its categories are broad enough to include just about everything. What's more, it has the advantage of offering an explanatory schema as a bonus. But it leaves the scientist unsatisfied, for it offers little leverage on the phenomena. When we examine the details of particular illusions—say, the contrast illusions (see chapter 6) or the geometrical illusions (chapter 12)—this classificatory schema sheds no light. Figure 4-3 provides a counterexample to James's ideas. When we make mistakes about familiar objects or about photographs of familiar objects, it is not for want of recognizing them. But some feature of the image suggests a surface property (convexity, sense of relief, closure, change of slope) that does not accord with those expected of the familiar object, and this feature, despite all indications to the contrary, creates conditions favorable to changing the interpretation of certain areas of the image, as in Figure 4–2, or even to a profound reorganization, as in

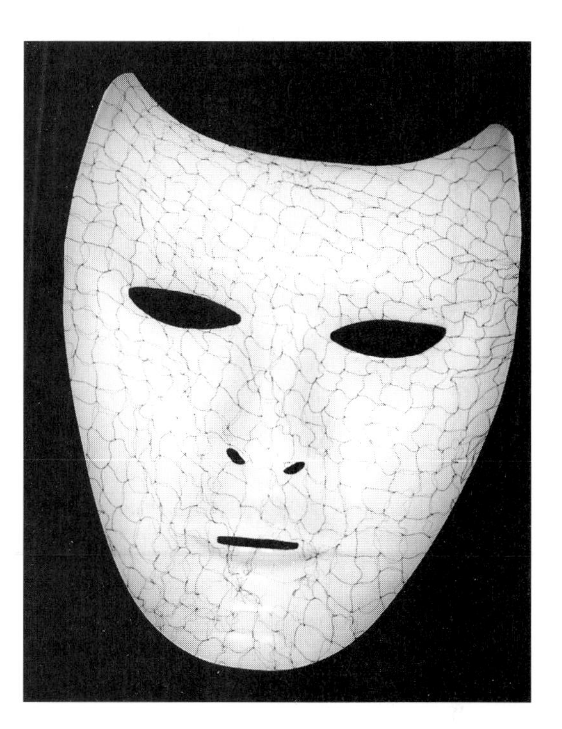

FIGURE 4-1. The illusion of the hollow mask. This theatrical mask was photographed from the back, but it is interpreted as if seen from the front. The net seems to fit the surface of the mask, whereas it was stretched from one edge to the other.

FIGURE 4-2. Two white walls seem to form an angle in the center, where the pole supports the television antenna. In fact, there is only one flat wall. When one turns this image around, one tends to interpret it as a photograph right side up, with roofs that would naturally top the houses. Photograph by Gérard Bouhot, seen in *Phot Argus* in 1994.

Figure 4-3. This behavior of the human brain, generator of illusion, seems to me scientifically impeccable. Many advances in science have been triggered by a small discordant result that might have been buried but ends up leading to the complete reorganization of a field of knowledge.

Until now, the classifications of illusions produced by scientists have been mainly catalogues of visual illusions, grouped according to the attribute on which the error bears: size, form, distance, movement. Richard Gregory has proposed an interesting double-entry system of classification that is both philosophical because of the very broad choice of categories and technical because of the choice of examples (Table 1). On the one side, the illusions are listed according to their effects into horizontal categories: "ambiguities," "distortions," "paradoxes," and "fictions." On the other side, they are divided into three vertical categories: “physical,” “physiological,” and “cognitive.” We have seen to what point the distinction between “physical” and “nonphysical” depended, in the past, on the state of knowledge. Much of the uncertainty about where to place this dividing line has been resolved. On the other hand, the border between the “physiological” and the “cognitive” is the subject of fierce debates, and Gregory, even if he is right, will be hotly contested on this point by specialists. But it is precisely the merit of this classification to dot the i’s and oblige us to take a stand.

Gregory has brought a very personal perspective to several illusions, notably Pulfrich’s pendulum (Figure 5-6), the chevron illusion (Figure 10–1), Münsterburg’s illusion (Figure 4-4 and Figure 4-5), the illusory contours (chapter 8), the blind spot (see Figure 2-1), and the Leviant il-

FIGURE 4-3 The lightbulb can be seen as either inside the lamp or outside it. In the latter case, the lamp’s surface looks domed and the ceiling fixture from which the lamp is hanging seems to come through the small opening, bringing the bulb in front of the fixture.

TABLE 1 Classification of illusions according to Richard Gregory

	Origin		
Type of illusion	**Physical**	**Physiological**	**Cognitive**
Ambiguities	Hazes	Ames room	Necker cube
	Shadows	Apparent motion	Figure-ground reversals
Distortions	Stroboscopy	Geometric adaptations	Geometric illusions
	Bending of light rays	Café wall (Münsterberg)	
		Contrast effects	
Paradoxes	Mirrors	Discrepancies between channels	Impossible figures
		Motion aftereffects	
		Constancies	
Fictions	Rainbows	Afterimages	Subjective contours
	Moirés	Phosphenes	Filling in the blind spot

lusion (Plate 1). However, he is being cited mainly for his analysis of the geometrical illusions, which is highly valued by philosophers and little supported by specialists. He himself seems very attached to this aspect of his work.

In Gregory's classification the horizontal categories are appealing in that they can be used for classifying the illusions of politics, love, or money as well as the oddities of language. Moreover, he has played at illustrating each category with verbal prototypes. Here are some personal adaptations.

Ambiguity:

(An episode from the French Revolution.) "We are here by the will of the people and will leave in no other way than by the power of the bayonets." This pround declaration, taken literally, could mean "Go and fetch a corporal and two soldiers with their bayonets, and we will hasten to get out."

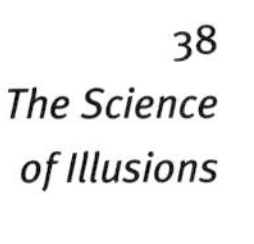

FIGURE 4-4. Illusions discovered by Münsterberg, around 1890. The bricks are exactly aligned horizontally, but the lines of mortar separating them seem to slant alternately up and down. It is important for the illusion that the successive layers of bricks be separated by the gray bands.

Paradox:

From a mother to her daughter: "Think twice before doing something silly"; or the daughter's answer: "You'll have to kill me to make me change my mind."

Distortion:

"My kingdom for a goat." (Richard II) His successor, Richard III, later proposed a more balanced exchange for a horse.

FIGURE 4-5. Tilings. Despite the regularity and simplicity of these designs, which are related to the chevrons illusion, observers make errors in judging the orientation of the columns—all of them vertical. The upper motif is borrowed from Nicholas Wade's book *The Art and Science of Visual Illusions*. The lower one is a variant of a collection by Sakusi, displayed on a Japanese web site on illusions.

Fiction:

"On the planet Mars, the citizens have the curious tradition of appointing an animal for president, and this year they chose a baboon."

Here I adopt a different plan. I do not think it useful to make a sharp distinction between the physiological and the cognitive. The information flows in both directions; taken in by the receptive organs, it is sent to and interpreted in the deep areas of the brain, which send signals back to the receiving organs, to refine their working conditions. Depending on their temperament, scientists favor one of the directions for propagating the information, and want to explain everything by the physiological or by the cognitive. The interesting idea of Gregory's that I have retained—but which is not essential to his classification—is that an illusion can be the consequence of a general procedure for information processing in the service of the individual. Gregory is right in particular to stress that perceptual constancy—the tendency to perceive an object as having its permanent properties rather than its changeable features—is the source of numerous illusions.

I shall thus classify illusions according to some of the main perceptual processes, be they auditory, visual, or other. In the conglomerate of acoustic or visual impressions, perception must first demarcate the borders that will enable the perceiver to separate from one another the things he sees or hears. This is the domain of contrast effects in the determination of boundaries (chapter 6), and that of segregation or fusion effects in boundary groupings (chapter 7). If need be, perception makes bets where the information seems incomplete, and makes us see or hear as if there had been no gap in the signals (chapter 8). Once a thing has been picked out and its limits defined, it must be assigned a size and a location. This implies the task of calibrating signals, which results in more or less rapid adaptations (chapter 9). As much as possible, perception provides a stable description of the object, which enables us to memorize and recognize it despite the many appearances it might have under different conditions; this is the domain of perceptual constancy (chapter 10). Finally, the object must be located in space, either absolutely or in relation to certain reference points (chapter 11). The signals received are processed in parallel by several modules. Perception must harmonize the sometimes discordant results or make arbitra-

TABLE 2 Classification of illusions according to the procedures involved.

Category	*Visual (V) and auditory (A) prototypes*	*Gregory's classification*
Perceptual limits	V: Benham's disc	P/fictions
	A: Seashell sound	idem
Contrasts	V: Mach bands	P/fictions
	A: Rawdon-Smith effect	P/distortion
Segregations, fusions	V: Ouchi's illusion	C/ambiguities
	A: Melody segregation	C/ambiguities
Completions	V: Subjective contours	C/fictions
	A: Auditory restoration	C/fictions
Adaptations	V: McCollough effect	P/fictions
	A: Zwicker effect	P/fictions
Constancies	V: Color of the moon	P/distortion
	A: Phase neglect	C/paradoxes
Reference points, localizations	V: The pigeon's head	C/paradoxes
	A: Deutsch's illusion	P/paradoxes
Arbitrations between channels	V: Geometric illusions	C/distortions
	A: Dominance of the image	C/paradoxes

All these effects are presented in the main text of this book with the exception of the Rawdon-Smith effect, which is explained in note 6. As regards the classification into "cognitive" (C) or "physiological" (P), the assignment often seems to me uncertain.

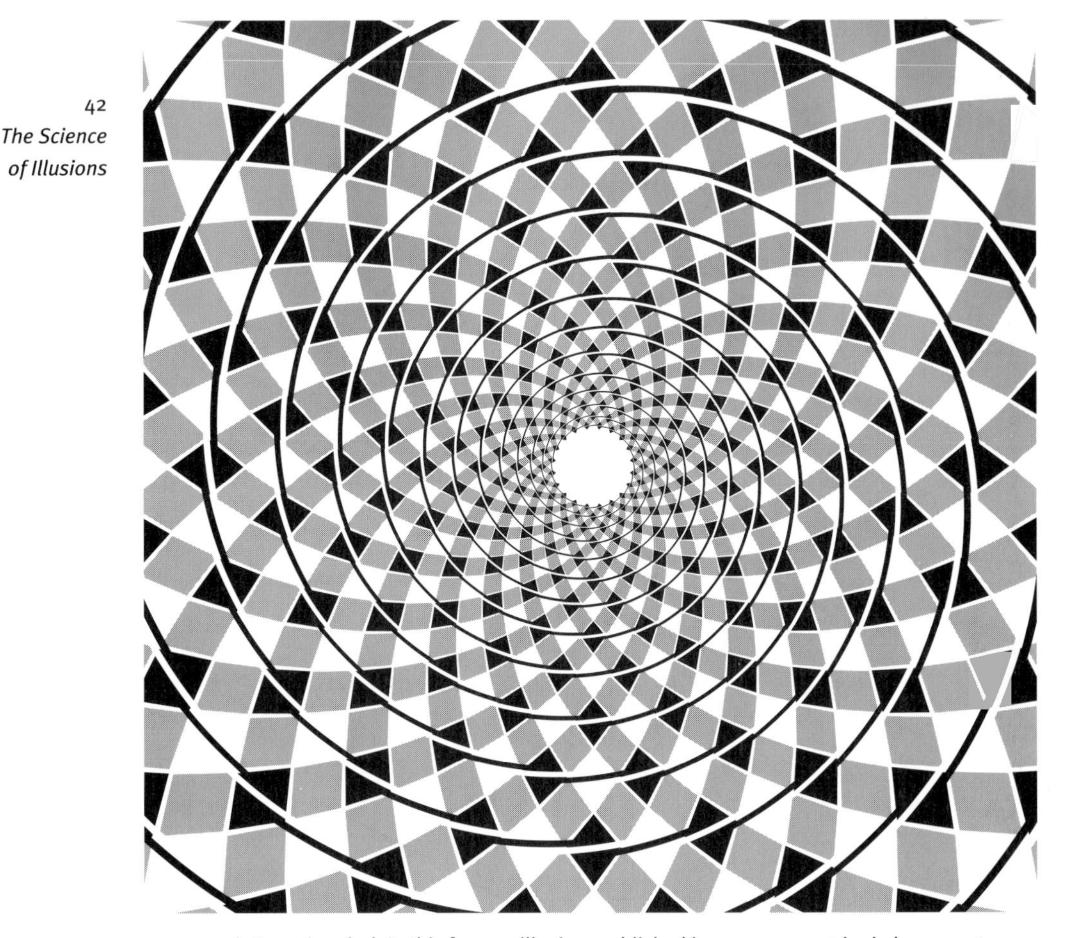

FIGURE 4-6. Fraser's spiral. In this famous illusion, published in 1905, concentric circles seem to form a spiral. The tiling is not necessary for the illusion, and the illusion is even stronger with thicker circles. Incidentally, note that one is sensitive to the designs in arcs of circles formed by the bands where gray and black squares or gray and white squares alternate. However, it is rather difficult to perceive the perfect alignments formed, in the direction of the center, by either the black squares all by themselves or the white squares all by themselves.

tions. The geometrical illusions are, in my opinion, related to this necessity (chapter 12). Each of these tasks produces effects that sometimes surprise us and are, in those cases, designated illusions. The scientist is interested in them because he or she hopes to extract from them the logic of the normal functioning of perception. Illusions that are attributable to real defects in the senses are fairly rare; they are generally due to the relative slowness of the work of interpretation (chapter 5).

These categories do not exhaust the subject. I would like to add to it the "illusions of illusions"—bizarre perceptual phenomena reported by several writers in support of an idea that is dear to them but that proves unfounded. Here then are some illusions that, as far as I can tell, are just illusions of illusions:

"The yellow we see in the rainbow is not really there, but an appearance resulting from the juxtaposition of red and green." (Aristotle)

> Every smooth surface, he [Aristotle] says, reflects rays coming from the eyes. Water and air have an even smoother surface than any other object. So if the air is dense, our gaze comes back to our eyes. When one's sight is poor and without power of penetration, whatever layer of air it encounters, it is powerless. This is the cause of a disorder from which some sick persons suffer: they see their own image everywhere and imagine they are going to encounter themselves. Where does this come from? From the fact that their sight is poor and that it bounces back, unable to penetrate even the nearest layer of air. (Seneca, adapted from Aristotle)

"If, on the other hand, we see the lightning before we hear the thunder, this is because vision is faster and arrives far ahead of audition." (Seneca) This explanation is incorrect, but it does contain a kernel of truth: one sense may inform us more quickly than another. This is what happens with the illusion of the touch telephone: your telephone call seems to go out before you have finished dialing the number because your auditory signal is treated more quickly than the tactile signal.

The wider-open the pupil of the eye, the larger objects look, said Leonardo da Vinci. But the observation of Plempius, quoted by Joseph Plateau, is probably more accurate: "If we look at a distant candle on a stormy night and in a place where the eyes can be caught by bolts of lightning, we see the candle's flame contract, and then dilate at each bolt. The reason is that each time the eyes are caught by the lightning, the pupil contracts and then dilates."

I also have doubts about this illusion as described by Montaigne: "Bodies we are looking at are perceived as longer and more stretched when we press on the eye." (Or perhaps I don't have the nerve to press as hard as he did.) In any case, he must have liked to manipulate his eyes. He also notes that if we press hard on the eye from underneath, things look double to us.

"Before children become acquainted, through touch, with the positions of things and of their bodies, they see the world upside down; thus they get from their eyes a wrong idea of the position of objects" (Buffon). The rest of this passage is more plausible, but unproved: "A second imperfection, which must lead children into another kind of error and false judgment, is that they first see all objects double because in each eye there forms an image of the same object."

"When one looks at a canal or a long lake, the horizontal plane of the surface of the water looks inclined, and seems to rise increasingly as its parts are farther away from the viewer" (Vergnaud). Is this not rather a statement about the way of representing a scene in perspective? This other observation by the same writer is probably more correct: "An inclined piece of land, seen from a certain distance, looks to us longer or shorter, depending on whether it is rising or falling, than when it is horizontal."

"When we look at ourselves in the bathroom mirror in the morning, we have an impression that our face in the mirror is exactly the same size as our 'real' face. However, if we trace its outline in the condensation, we realize that our reflection is in fact ridiculously small" (J. Mehler and E. Dupoux). In fact, the image of the face is not *on* the surface of the mirror, but behind it and has the same size as our real face. The kernel of truth in the experiment with condensation is that we do not need the whole surface of the mirror to see our whole face; it is enough for the mirror's dimensions to be half those of the face.

Finally, there is an infallible technique for creating false illusions in the laboratory: the technique of forced-choice alternatives (FCA). If I prick your skin with a needle, you will feel a single prick and not have the illusion of a double prick at two close points. Here is how to produce such an illusion in the laboratory: the experimenter is, say, interested in "tactile resolution"—the subject's ability to feel separately two extremely close stimulations on the skin. The density of the tactile receptors under the skin is limited; when two pricks are so close that a single receptor detects them, the subject feels only one. Sensitivity is acute on the fingertips (two millimeters), middling at the tip of the nose (seven millimeters), and poor on the neck behind the ear (thirty millimeters). The experiment is easy to do with someone blindfolded and using as an instrument the two points of a compass or two pins stuck through a cork.

To make a precise determination of the threshold at which one is capable of distinguishing between one prick and two, behaviorists introduced a method of FCA in which situations are systematically created where the subject finds it very hard to know what he is experiencing. He is thus pricked with either a single point of a compass or with two points but so close together that the subject has trouble distinguishing between one and two pricks. As predicted, the subject will often say that he feels one point whereas he was pricked by two. But also, knowing that two nearby pricks can be felt as one, he will frequently say that there were two pricks when there was only onc.

The FCA method, which measures essentially the subject's suggestibility, is one of the pillars of contemporary experimental psychology, and its results are well established. In daily life we know that when we ask someone a question that is too simple, he or she will often look for complications and will answer in a way that does not reflect what he really thinks. Thus by asking carefully chosen questions, we get people being polled to say just about anything we would like them to say.

LIMITS

Few illusions are due to an imperfection in the sense organs. These organs are extremely sharp in comparing subtly different sizes, and they have a remarkable sensitivity for detecting the faintest signals. At the most, we can criticize vision and audition for a poor assessment of absolute sizes and, for vision or touch, a certain slowness.

The quality of an optical instrument is judged first by its "resolving power"—its capacity for giving two separate images of two points located in adjacent positions. In the most precise region of the eye, the fovea, the cells for detecting light signals are separated by about three thousandths of a millimeter, which corresponds to an angular separation of about a hundredth of a degree (.17 millimeters at a distance of one meter). We are able to distinguish targets whose directions subtend an angle of a few seconds of arc—that is, ten times better than the separation of the detector cells. This "hyperacuity" is obtained thanks to an intelligent processing system that takes best advantage of the signals supplied by the eye's somewhat limited optics. At a distance, parallel fine lines merge into a gray area, the textural details become blurry, and a figure merges with the background (Figure 5-1).

Another problem with the eye is its mechanical instability: it is constantly engaged in tiny tremors, and when it changes direction, driven by the brain, its movements are not tremendously precise. This problem shows up in the Bourdon figure, where it is impossible to count the points, for we do not have sufficiently precise control of our gaze's jumps from one point to the next (Figure 5-2). In the sky the stars are seen with a characteristic twinkling. Some observers see five or six branches stretching or retracting in no

FIGURE 5-1. Carlson's figure. In this image created according to an idea of Carlson et al. (1980), the design shown fades when it is observed from a distance of about three meters. The principle is that the gray area is homogeneous; the prominent black marks are surrounded by white, which makes them stand out when seen close up, but cancels the black when seen from a distance.

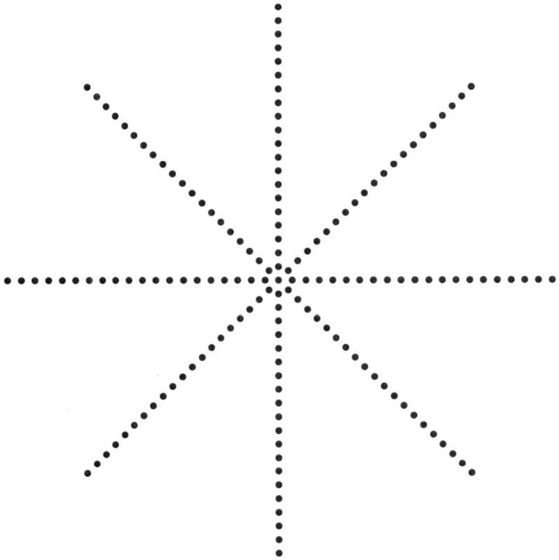

FIGURE 5-2. Bourdon's figure. It is practically impossible to count its points (forty per line), because of a lack of precision in guiding the movement of the eyes. Secondarily, one can compare the lengths of the lines. Their equality is perceived better when one directs attention to the angle formed by two adjacent segments.

order. This phenomenon, in my opinion, is more perceptual than it is physical. Because the light from stars or planets encounters temperature fluctuations in the layers of air it penetrates, its image undergoes jumps of the order of a second of arc. These movements have no great effect on the appearance of Venus, whose apparent diameter varies between three and ten minutes of arc. It is different with stars. Their apparent diameter is always less than a second of arc; observed with a powerful high-resolution telescope, they do not stop moving. But I have trouble making the connection between these fluctuations of position and the appearance of the star to the naked eye.

The major drawback of the visual apparatus is its slowness, which is the source of several effects that have been known for a long time. Aristotle gave the example of a burning ember that one turns, holding it at the end of one's outstretched arm, thus producing a complete luminous crown. Ptolemy described the fusion of colors in a spinning top. J. Plateau reported his experiments as follows:

> If a disc is painted various colors along directions going through the center, and if it revolves at high speed, it appears to have a single color, specifically, the color that would result from the mixture of those painted on the disk. The reason is that one's gaze does not remain on any of them but falls successively on all of them so that, because of the speed of this succession, the eye cannot distinguish the colors from one another; all the colors thus appear at the same time over the disc's whole surface, as if they were just a single color.

Ptolemy also proposed that the arcs made by shooting stars are an illusion, owing to the persistence of the sensation of light. Conversely, Ibn el-Haytham observed that certain objects that move too fast are not seen, that a wheel that is turning rapidly appears to be immobile. Later, others were to observe that a bullet or cannonball is not seen in its course, that the fast-spinning blades of a fan look transparent, that a vibrating rope appears to fill the whole space over which its vibrations extend.

In 1838 Gustav Theodor Fechner described a disc made up of six black areas of increasing sizes (Figure 5-3). When the disc revolves rapidly, we expect to see gray bands form, dark ones near the center, where there is more black than white, and light ones toward the periphery, where the white predominates. The curious thing is that what is formed are colored bands. Fechner proposed an ingenious explanation for this. The white of the disc sends

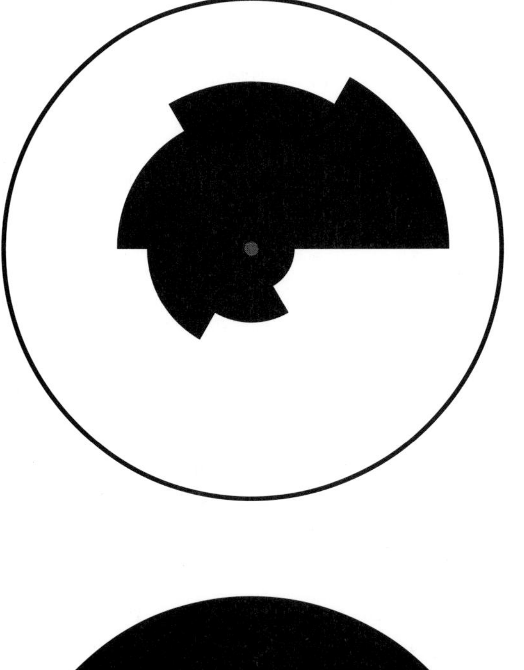

FIGURE 5-3. Subjective colors. When one of these discs is rotated about thirty revolutions a second, colored rings appear. The upper disc was described by Fechner in 1838. The one below, much more common, was marketed as a toy by Benham in 1894. When it turns clockwise, it forms, from the center to the periphery, red, green, pale blue, and dark purple bands. When the disc revolves counterclockwise, the same sequence of colors is formed, but from the periphery to the center.

back white light, which is composed of light of all colors. A receptor in the retina, sensitive to the color red, is inactivated when a black area passes, and the receptor takes time to respond next to the change of light. If the receptor for blue is slower to react, the brain receives, for a fraction of a second, a description of the composition of the light, a description biased in favor of red. If this explanation is correct, only the durations of the successive exposures of white and black count, and these durations do not depend on the disc's direction of rotation. However, the colors actually perceived do depend on the direction in which the disc revolves.

In 1894 a variant of this disc was marketed as a toy. One axis, going through the center, allowed the disc to spin like a top; this is Benham's disc (see Figure 5-3). In good light, the best effect is obtained at about five or ten revolutions per second. In addition, the colors depend on the direction of the rotation. Retaining the idea of the shifted response times, we favor the view that the subjective colors attributed to the disc follow from comparative analyses, the brain deciding the color of an area depending on the colors of the neighboring areas.

It is possible to make subjective colors appear at slow speeds: for this it is enough to use black and white images, subdivided into very fine, contrasting periodic areas. Slow movements are then enough to create fast alternations of black and white, which give rise to subjective colors (see Figure 11–6). When the divisions are even finer, it is no longer necessary to move the image. The instability of the gaze is enough to create apparent movement. The best-known figure is that of D. M. MacKay, in which a surface is divided into 240 very fine alternating black and white sectors, radiating from the center (Figure 5-4). When you stare at it for about ten seconds, you have a visual sensation of swaying movement and, if you pay attention, you see a revolving movement. Generally, the figures with dense thin lines produce motion effects perpendicular to the lines. In another variant, also proposed by MacKay, numerous segments form parallel stacks (Figure 5-5). Here, we see a "streaming," as if water flowed through the spaces separating the stacks.

This illusion, which was not compelling in MacKay's original figure, takes an extraordinary turn in the pictures of the "Enigma" series painted by Isia

FIGURE 5-4. MacKay's illusion. When we stare at this image for about ten seconds we have a sense of undulating movements that become revolving movements. Using 110 subjects, MacKay found eighty-three who saw the movement take a clockwise direction, and twenty-seven who saw it move in the reverse direction, with some subjects capable of alternating. To see the direction of rotation change, MacKay recommends fixating a point to the left or the right of the center of the design.

Leviant in the 1980s. These pictures combine the radial organization of MacKay's first figure with the principle of streaming in the open spaces of his third figure. In it we see revolving movements form in the rings interrupting the radial lines (Plate 1). Leviant achieved this result through a systematic search for the conditions that promote the illusion: (1) the thickness of the dark rays must be more than 50 percent greater than that of the light-colored rays; (2) the rays must be interrupted at a right angle; (3) the width of a band must be equal to 3.2 times the thickness of a pair of black and white rays; (4) the colors are not important, but the average luminance level of the rays must be equal to that of the bands; (5) several bands are more effective than a single one; (6) the bands must be perfectly homogeneous (all traces of brush strokes must be eliminated). Paradoxically, when the movement is perceived, we have the sense of a textured (grainy) surface spinning.

The source of the illusion is a matter of debate. For some people, it would be due to the mechanical instability of the eyes. But why, under these conditions, is a stable circular movement created? Leviant prefers to invoke deeper phenomena in which the neural circuits responsible for detecting motion could be influenced by other neural circuits at work nearby dedicated to the analysis of spatial orientations.

In audition, sound sensation results from the analysis of variations in air pressure in the outer ear. According to E. Leipp, we perceive vibrations whose amplitude is "weaker than the diameter of one molecule of hydrogen (10^{-8} millimeters)." When we place an ear next to the opening of a seashell, or close to a long pipe open at both ends, we hear "the sound of the ocean." This is not an illusion, but on the contrary, testimony to the sensitivity of the ear, which picks up infinitesimal fluctuations of pressure inside the seashell or pipe.

We attribute pitch to sound frequencies ranging from some 20 Hz (vibrations per second) to 15–20 kHz. Under 20 Hz, the sound signals are perceived as discontinuous, or are not perceived at all: periodic variations of pressure on the order of a second do not produce any auditory sensation in humans. The weak sensitivity to low frequencies makes it possible for us to be unaware of many bodily noises, and not to be deafened by the beating of our hearts. Above 20 kHz, it is ultrasounds that are silent for us (although

FIGURE 5-5. Streaming. When we look at this image long enough, we have a sensation of flowing in the channels separating the blocks of parallel lines. A more striking form of this illusion where the streaming follows circular channels was discovered by Isia Leviant. (See Plate 1.)

other animals detect them and make use of them). Hence the following paradox: if one sound is added to another sound such that the sum of the two makes an ultrasound (which in principle is always possible), we should hear neither the first sound nor the second.

Nearly always, the senses provide excellent relative judgments; they are made to compare and distinguish. In audition, we are capable of detecting differences of 1/300th of an octave between successive sounds, and variations in intensity of two or three decibels. On the other hand, the ear is less reliable in the absolute. One notable exception is that of "absolute pitch" enjoyed by certain professional musicians, which enables them to know, when some isolated note is played, whether it is C or the D just above it. Having absolute pitch may be felt a handicap. One baritone complained that the perception of the notes in all their individuality was incompatible with the perception of the melody: the part did harm to the whole, as if in one of Giuseppe Arcimboldo's paintings we saw only the pieces of fruit and were unable to see the face they make up.

One hears one's own voice through both the vibrations transmitted in the air from the mouth to the ear, and through internal vibrations that only the person speaking can hear and that are transmitted by the bones of the spinal column and the jaws. This internal component of the voice is not recorded in sound recordings; hence the listener has the impression that he has a reedy voice with little timbre. Perhaps the art of singing rests on a deception: the person singing does his best to produce a sound that he finds pleasant, and people enjoy it because they are hearing something else.

The tennis or Ping-Pong player can sometimes be surprised by a "sliced" ball whose trajectory seems contrary to the laws of mechanics. The changes of direction of sliced balls were explained by Newton: the air pressure is not equal on both sides of a spinning ball, from which comes a trajectory that curves. For a few years Japanese volleyball players wreaked havoc with "floating serves" in which the ball changed direction during its trajectory. Here the explanation is more complicated than Newton's; it involves the turbulence of the streams of air through which the ball makes its way. There is, however, a curious, almost certainly illusory, effect encountered by baseball players. At one stage of the game, the pitcher throws a low ball to the batter. The pitcher must make batting hard for the batter by pitching him balls as hard to hit as possible within the limits of the regulation space. The two players are twenty meters apart, and the ball travels very fast. The bat is slender and thus must be handled with great precision to have a chance of hitting the ball, whose trajectory must be gauged in a very short time. When the ball is fast, the batter who has seen it start downward often has the illusion of seeing it rise at the end of the trajectory.

According to M. K. McBeath, the player underestimates the ball's initial speed. He thinks it is farther away than it really is; and when it comes near him, below eye level, he thinks it is lower than it really is. Then it appears to "catch up" in the final split second, which makes the batter see the ball rising at the end.

The same kind of reasoning applies to a much older and classic illusion, that of "Pulfrich's pendulum," which is often displayed in museums of science. The onlookers are invited to observe, through two lenses fixed to a partition, a pendulum that swings in a frontal plane a yard or two in front of them. One of the two lenses is strongly tinted and absorbs about 90 percent of the light. The pendulum must be observed with one eye in front of each lens. Lucky people will see the end of the pendulum travel through an elliptical path (Figure 5-6). This illusion is reserved for those who have stereoscopic vision (80 to 90 percent of all people do) and who moreover are not much bothered by the fact that a lot more light enters one eye than the other. Finally, another condition for the success of the experiment is that the pendulum moves about in a rich visual environment, so that its trajectory is judged in relation to reference points, a detail generally forgotten in museums.

The classic interpretation of this illusion is that the much-reduced light received by the one eye is analyzed more slowly than the light received by the other eye. Under these conditions, when the brain treats the signals coming

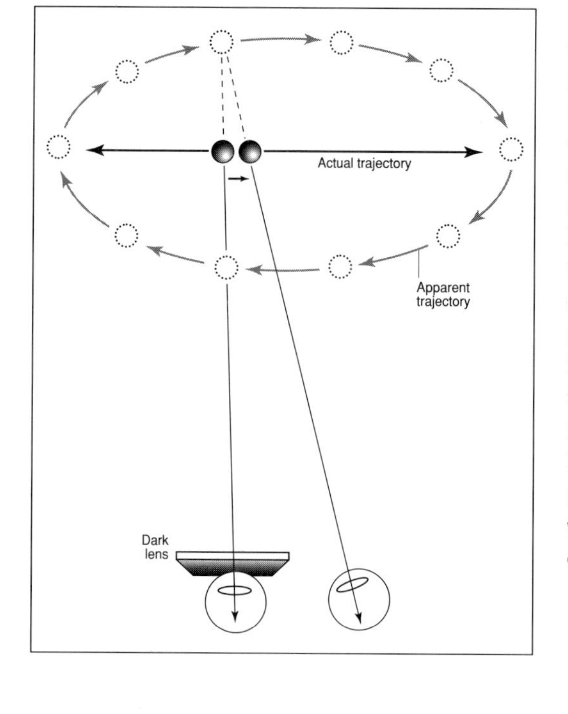

FIGURE 5-6. Pulfrich's pendulum. An observer tracks the movement of a pendulum that swings in a vertical plane, and a dark lens placed in front of his left eye. The attenuated image received by this eye is analyzed more slowly than the normal image received in the right eye. In stereoscopic vision, the brain combines the right-eye image of the end of the pendulum at a given point in its trajectory, with the left-eye image of this end at a point where it was a fraction of a second earlier. The intersection of the left and right sight lines, offset in time, gives the apparent position of the end of the pendulum, which seems to move along an elliptical trajectory.

from the pendulum at point A via the left eye, it is already treating the signal coming, via the right eye, from the pendulum at point B, where it is a split second later. Its apparent position is then at a point C, the meeting point of the two sight lines, that of the left eye toward point A and that of the right eye toward point B. This point of intersection is back of the plane of swinging when the pendulum swings from left to right, and in front of this plane for the return path.

A related effect is, it seems, observable when we are riding in a car. You must hold up in front of one eye a dark lens (for example, a lens of some good sunglasses). If you cover the right eye this way and look at the scene through the right-hand window, the car seems to go slower, and objects of known sizes look miniaturized. It you look out the left-hand window, you experience the opposite effects. Jim Enright, who observed these effects in himself and his family, said the effects are very pronounced at fifty miles per hour and are already perceptible at twenty-five miles per hour. Moreover, it is important that the environment be visually rich and not too bright; a road going through a forest is ideal for this. The explanation would be, as for Pulfrich's pendulum, that the right eye sees an object where it was a fraction of a second before reaching the position where it is seen by the other eye. The stereoscopic combination of the two pieces of information would lead to the illusions of speed and size.

6 CONTRASTS

We evolve in a world where everything is measured in relative values, in deviations from a norm. For example, in a place where an adult's average height was five feet eight inches (one meter seventy), a man shorter by 20 percent (thus measuring four feet six inches [one meter thirty-six]) would be judged a dwarf, and another man taller by 20 percent (measuring six feet eight inches [two meters four]) would be considered a giant. Decisions to buy are connected with small differences in price, lower or higher relative to what we think is the happy medium: a bit above it and we find the prices intolerably high and we do without some coveted item; a bit lower and we buy the article offered without needing it. Customers are ready to accept all kinds of constraints in order to economize on small amounts of money without realizing that such economies are canceled out by the slightest slip-up about a large sum.

Without our knowing it, perception forces this way of appraising things on us. Over moderate ranges, it sets up dividing lines where a signal varies continuously, to separate the strong from the weak, and in the extreme regions, things are treated without subtlety. For example, observe the beams produced by spotlights scanning the sky. At a certain distance the beams appear to stop, whereas their intensity changes continuously from the starting point to infinity; perception has placed a boundary that limits their length for us. It is true that a physical component enters into the phenomenon. We see the beam indirectly, for it is diffused by the atmosphere; the dust helps us to see it (as in a room, the dust helps to materialize the light rays). Because the density of the dust changes with altitude, the quantity of light diffused undergoes a marked drop at a certain point. Nevertheless, in this case a continuous change is transformed into an all-or-none.

A phenomenon of this kind was noted as early as 1769 by the abbé Diquemare. Observing the tail of a comet, he first noticed that the more he looked at it attentively, the longer it looked. Then, when he interposed his hand so as to see only the end of it, the end completely disappeared. Moreover, hiding the first third of the tail caused the other two-thirds to vanish. Here too, the light changed continuously, and perception rather arbitrarily set up a frontier. For Diquemare, the illusion lay in the tendency to see the comet's tail beyond its natural border. I am not sure that he was right about this point; the illusion could be in the elimination of the final third when the rest was hidden.

In any case, an observation of Rudolph Arnheim, made under more familiar conditions, is in line with the second interpretation:

> Our black cat lay on the windowsill, against the black night outside. When his eyes were open, his body was visible, dimly outlined; but as soon as he closed them, the whole cat vanished, leaving only the unremitting darkness of the window.

In the domain of acoustics, there is perhaps a phenomenon similar to the one described by Diquemare. Often, when I am listening to music at home, I have a sense that it is being played louder and louder, and I must turn the volume down on the speakers repeatedly. I also remember that once during a train trip, while I wanted to concentrate on some work, I was bothered by a conversation going on in a foreign language. As time went on, I heard the voices growing louder and louder, as if all the other passengers in the compartment were simultaneously raising their voices. A conversation is bothersome when one understands it. Here the voices became more and more familiar, characteristic, and perhaps because I heard them better, I also heard them as louder.

Objects look a bit larger when they are well lit. This is the "irradiation" effect, shown in Figure 6-1, where the white flagstones against a black back-

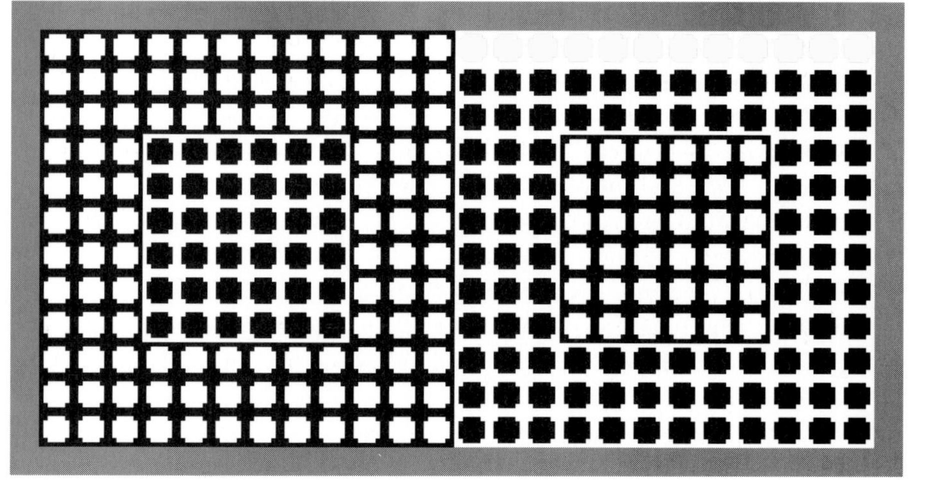

FIGURE 6-1. Irradiation. The white surfaces look larger than the black ones. We can notice this for the flagstones, which are all the same size, for the "alleys" between the flagstones, which are equally thick, and for the squares formed in the crossing of the alleys, which are all equal. Incidentally, we can notice an illusion about the estimation of surfaces: the flagstones seem to take up more surface than the alleys, but each exactly counterbalances the other. More generally, we tend to underestimate the surfaces of elongated figures.

ground look larger than the black ones of the same size against a white background. Another example: when the moon shows a well-lit crescent, and the rest of its disc is in shadow but visible, we notice that the disc looks larger on the side of the crescent. A stick held in the darkness in such a way as to cut off a source of light (such as the crescent of the moon, a candle flame, or lamp) shows a constriction at the point of intersection: the light source spreads out at the cost of the stick and thins it. Fashion magazines advise women on the heavy side to wear black dresses or black stockings, which are said to slim the figure, and black gloves in order to have small hands. But why is the black dress not also said to look shorter?

Irradiation raised serious problems for astronomers. Tycho Brahé observed that during solar eclipses the dark disc of the moon passing before the fiery disc of the sun appeared to lose about a fifth of its diameter. The most disturbing aspect of irradiation is that it is misleading at the very moment when the two discs begin to "come into contact" with each other and thus creates difficulty in measuring the crossing time of one disc by another, from which follows the estimation of important astronomical parameters. That is why Kepler, at the same time as he was trying to understand the formation of images in the eye, attempted to separate the possible physical causes from the subjective causes in the effects of irradiation.

If one picks out a star just at the moment when it appears to go behind the lunar disc, and one then focuses on it with an astronomical telescope, one again sees the star separated from the moon "by a space of one or two fingers," Gassendi tells us. An even more paradoxical observation is the one made by La Hire, during the passage of a star in the constellation Taurus behind the moon: at the start of the eclipse, he had been able to see the star inside and thus *in front of* the lunar disc. The interpretation is that the star was still alongside but the lunar disc was seen as too large, so the star was seen as inside the disc. This is a remarkable phenomenon, and I do not know of any laboratory experiment that illustrates it.

The "Mach bands," described and analyzed by the Austrian philosopher Ernst Mach around 1860, are illusory bands, light or dark, that we see on the border between shadow and penumbra. If an object cuts off a light flow and we observe in detail the border between the shadow projected on a surface and the illuminated part, we see two bands on each side of the border, a bright band in the light zone and a dark band in the shadow zone, which together accentuate the differences in light intensity (Figure 6-2 and 6-3).

As early as 1865 Mach proposed a neural model to account for the phenomenon, a model that was to prove essentially correct. He proposed that the neurons transmitting the measures of light intensity beyond the retina interact laterally: one neuron inhibits its neighbors (that is, causes them to transmit weaker values than the ones they would have signaled). The closer the neuron's neighbors, and the higher the signal level it transmits, the greater the inhibition.

FIGURE 6-2. Mach bands. This figure contains a light gray area to the left, a dark gray area to the right, and a shading in the middle that completes the transition between the two areas. We see a light vertical band on the left and a dark vertical band on the right of this gradation; both of them are illusory.

FIGURE 6-3. The Mach-bands illusion in astronomy? The picture represents the appearance of the planet Saturn emerging from the lunar disc after occultation. The moon's luminosity is in fact much greater than Saturn's, and a dark band appears at the border between the two. In the beginning, it was seen as the sign of the presence of a gaseous atmosphere on the moon. Adapted from Lucien Rudaux, *Sur les autres mondes*, p. 58.

The Mach-bands illusion occurs at an early stage: it happens immediately after the retina. The information transmitted to the brain has a characteristic "signature" with two effects. One is the effect described, two illusory bands. The other one, which corresponds to the normal role, is to indicate the existence of two adjoining areas of different luminosities.

This function is the source of an artificial illusion, that of Craik-O'Brien or that of Cornsweet. If we take a gray area of uniform luminance, and divide it into two zones by a border made up of two very thin juxtaposed bands, one light and one dark, the two regions thus delimited will appear to have different luminosities. The side where the dark band is will look darker than the side with the light band. This effect, which is clearly seen with revolving discs, is hard to reproduce in a photograph. We can get an idea of the effect produced with the diamonds of Figure 6-4 or the variant on Plate 5.

Rather than drawing up a point-by-point map of the light intensity received, the retina apparently transmits to the brain condensed information in which figure the borders between clear and dark regions and the average brightness values on both sides of the borders. One illusion, the "Hermann's grid" (Figure 6-5), is a case in point. In this grid, the regularly arranged black squares are separated by thin white bands. At the crossing of these bands illusory small gray spots appear. To understand them, imagine neurons that transmit measures of light intensity at various points on the white bands. Let us suppose that instead of sending absolute measurements, the neurons send somewhat corrected measures, taking account of the average intensity in the vicinity of the point of measurement. More precisely, imagine that the neurons treat as lighter a point surrounded by a dark region, and as less light a point surrounded by a light region. The points found in the intersection of the white lines are those where the environment is lighter, and hence the application of a correction that makes them look gray.

The study of various versions of the illusion (Figure 6-6 and Figure 6-7) brings to light other phenomena that are not entirely consistent with this explanation. The effect of the grid is not local: it is seen provided that the grid has numerous elements, and when one stares at an intersection, the effect is primarily seen farther away on the grid, from the corner of the eye. We

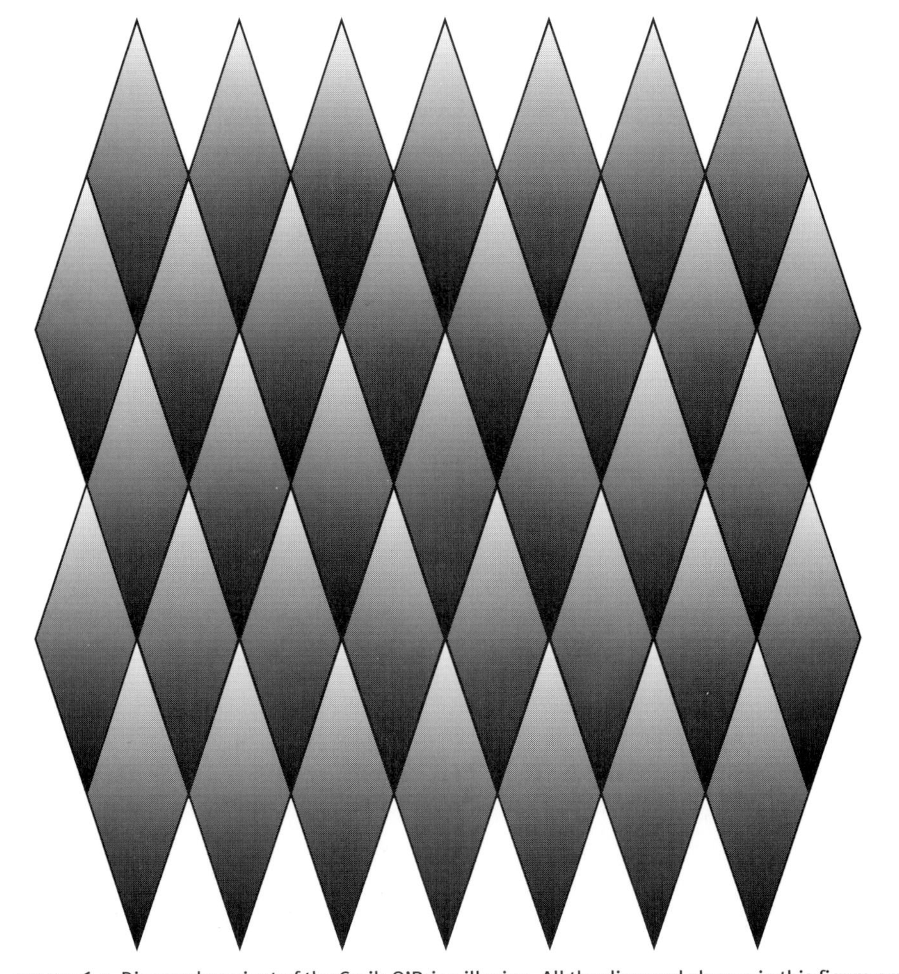

FIGURE 6-4. Diamonds variant of the Craik-O'Brien illusion. All the diamond shapes in this figure are identical, but the lower rows look lighter than the upper ones—it is an illusion. In the Mach bands, the transition between two areas of different shades of gray was accentuated by illusory pale and dark bands. Here we have the reciprocal phenomenon in which a rapid change from the light to the dark on both sides of a border is propagated and creates gray levels, the averages of which differ on the two sides of the border. The real Craik-O'Brien illusion, which is very unconvincing on paper, is not shown here. The very effective variant with diamond shapes was described for the first time by Watanabe in 1995.

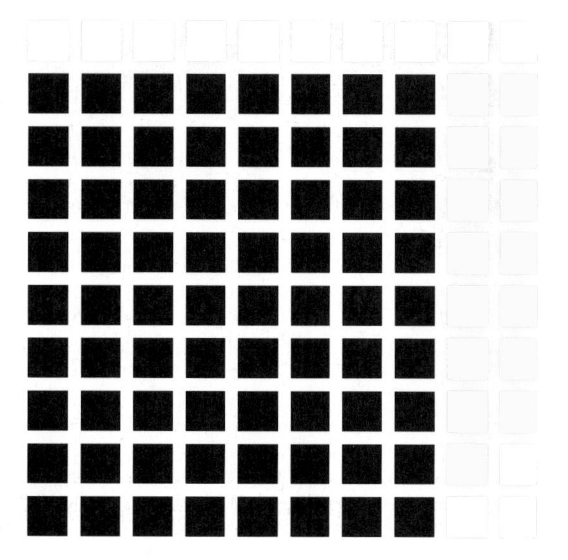

FIGURE 6-5. Hermann's grid. Small illusory gray spots appear in the intersections of the white bands. The spots are better seen "from the corner of the eye," in the periphery of the areas one stares at.

FIGURE 6-6. The figure on the left is a Hermann's grid. In it we find illusory gray spots. When we observe it after rotating the page forty-five degrees, we see two networks of dark lines go through it according to the diagonals of the quadrilaterals. The same kind of line is observed in the figure on the right, which is of the "compressed checkerboard" kind.

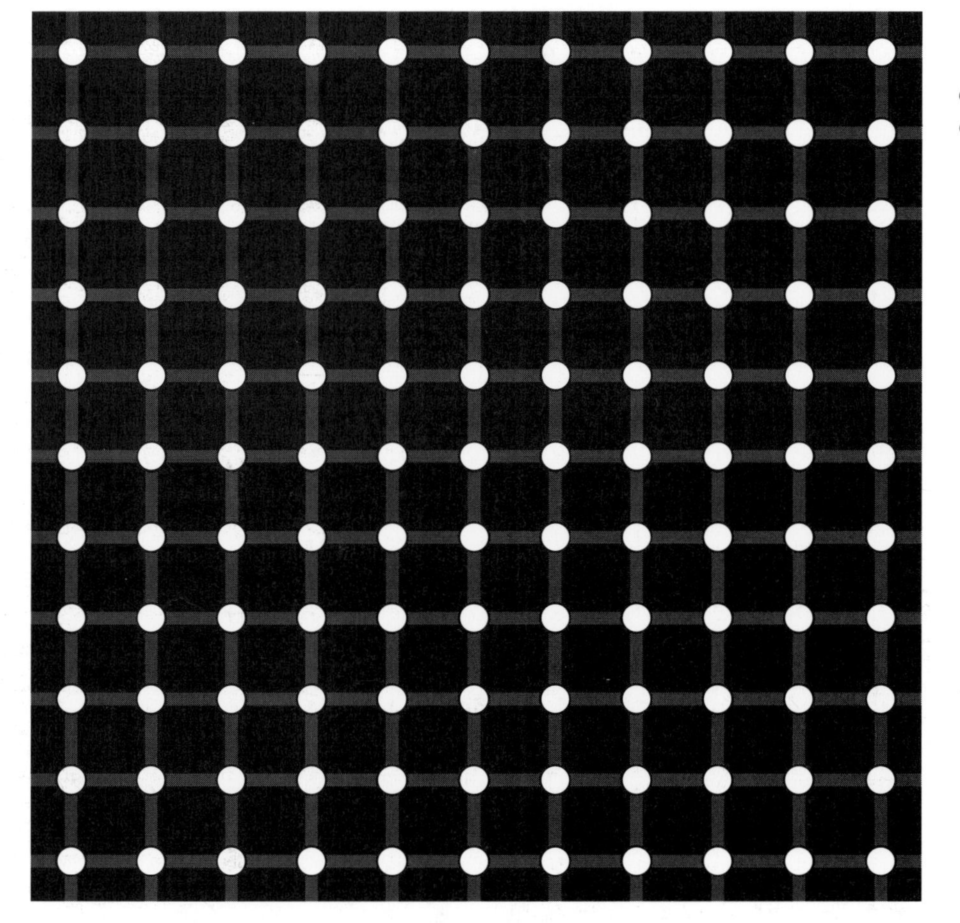

FIGURE 6-7. Scintillating Hermann's grid. Described by Schrauf in 1997, this grid scintillates when we look at it absentmindedly: dark spots appear inside the white discs when we move our gaze. The black circles around the discs are there for decoration.

would expect there to be a close relation between the geometry of the grid and the arrangement of the receptors on the retina, which is no more "square" than it is hexagonal. There is no effect with a hexagonal grid, but there is one when the square grids have gone through various distortions (for example, Figure 6-6), provided they do not interrupt the alignments of the sides of the elements. When we turn the grid forty-five degrees, the gray spots disappear and the salient phenomenon is the observation of dark lines going diagonally across the elements. We begin to wonder whether the phenomenon is retinal or occurs much later.

Painters have long known that, in a picture, the appearance of a color depends on the colors nearby and that the luminosities there influence one another. In the assimilation effect (Figure 6-8), a gray background crossed by fine, dense lines takes on a value similar to that of the lines. It is pulled toward the white when the lines are white, and toward the black when they are black. When larger areas are in play, the reverse effect, one of contrast, is observed (Figure 6-9). A situation was recently discovered in which juxtapositions of large surfaces diminish the contrast: this is the Munker-White illusion (Figure 6-10, and Figure 6-11). In Figure 6-10, it is the gray lined on

FIGURE 6-8. Assimilation. The gray of the center square (between black lines) looks darker than the gray that surrounds it (between the white lines). In fact, the same gray was used over the whole surface of the picture. This consistency can be confirmed by hiding the ends of the white or black lines near one of the corners of the small square.

FIGURE 6-9. An effect of simple contrast. Both discs are the same shade of gray, but the one on the right, on a light-colored background, looks darker than the one on the left, on a dark background. The presence of backgrounds with various shades of gray intensifies the illusion.

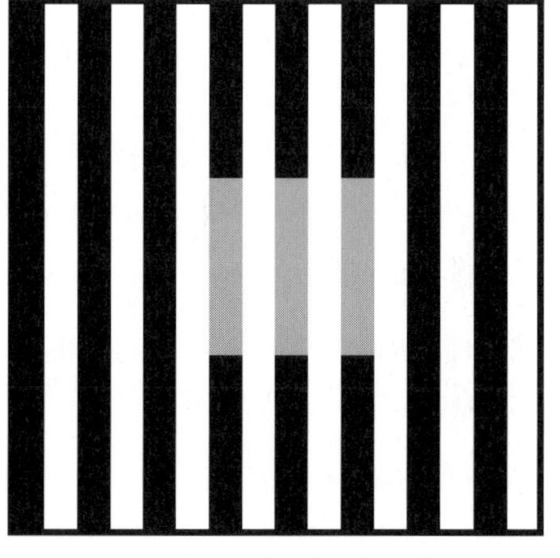

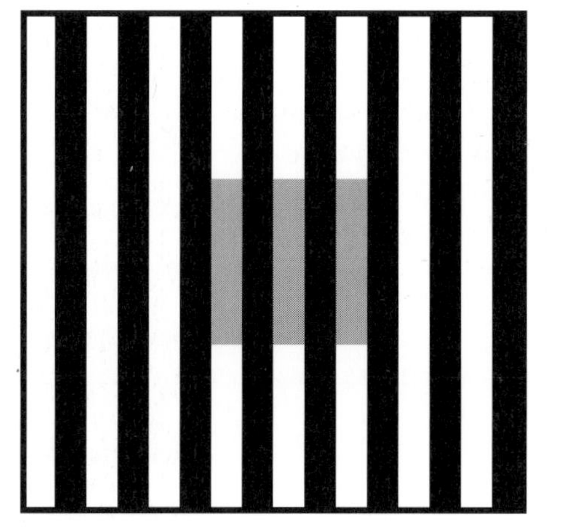

FIGURE 6-10. Munker-White illusion. The gray square on the black stripes (left) is the same as the gray square between the black stripes (right), but it looks much lighter.

both sides by black bars that looks darker than the same gray lined by white bars. Assimilation occurs in the transverse direction, and contrast in the longitudinal direction.

In a moderately lit room a television screen looks gray. When the set is turned on and a so-called black-and-white film is playing, the darkest parts of the image become plainly black whereas the beam that creates the images on the screen can only add light and never take it away. The gray of the unlit parts of the screen has become black, by contrast with the white parts.

There is, I believe, a subtle logic involved in seeing a beautiful black hue and a striking white hue on the television screen. Imagine I have laundry to hang on the rungs of a ladder propped up against a wall. If I have two items of clothing, I shall put them on the rungs within easy reach. I shall tend to separate the two items, but still leave them at heights that are not too different. If I have more items, I shall tend to distribute them over all the rungs and thus, when there are a great many of them, to take up the highest rung and the lowest rung of the ladder. By analogy, if the screen contains many

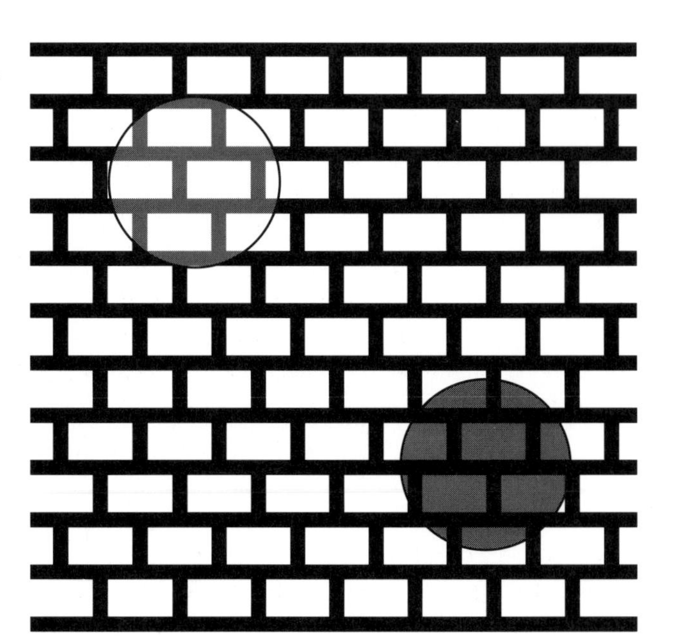

FIGURE 6-11. Variant of the Munker-White illusion, with an effect of transparency on the upper left disc. The same gray was used for both discs, but the one on the left looks lighter. The black outlines of the discs do not add to the illusion.

shades of gray distinguishable from each other, I shall tend to assign the "black" value to the darkest one and the "white" value to the lightest one. If the screen was simply split in two and each of the two grays entirely covered one of the halves, I would actually see two gray areas on the screen.

How many categories do our perceptual scales contain? About seven. We are able to detect infinitesimal differences between two stimuli. Here the word "infinitesimal" is as important as the figure "two." This discriminatory power holds only for the sensations being compared two by two at the very moment they are compared. If we try to memorize several stimuli, and then seek to classify a new stimulus among the old ones, we are unable to do so precisely; it is as if what had actually been perceived had been distributed among five to nine categories.

What is the whiteness of snow? Leonardo da Vinci noted that "the snow that falls looks dark when we see it against the sky, and bright white when we see it against the dark background of the window of a house." Another paradox, observable in overcast weather on ski slopes: the sky is gloomy and the snow looks very white although, getting its light from the sky, it cannot be brighter than the sky. According to the analysis done by Koenderink and Richards, this would not be an illusion. We look at the snow at our feet, which reflects the light coming from above. On the other hand, we appraise the darkness of the sky by looking far in front of us. The diffused light received in this direction, close to the horizontal, would really be less intense than the light received and sent back by the snow in a direction close to the vertical.

This explanation is reminiscent of the hypothesis advanced a few years earlier by Pomerantz to explain why the neighbor's grass is always greener. According to Pomerantz, the gardener, next to his fence, sees at his feet some sparse grass against a background of brown earth. But he sees his neighbor's grass, beyond the fence, from a more horizontal angle. From there his eye receives some green coming from several thicknesses of blades of grass. He has little chance of catching sight of the earth from this angle, and the green of the neighbor's grass is thus appreciated in all its intensity.

Another paradox known to skiers is that the wearing of yellow lenses in bad weather provides a sense of increased brightness. Nevertheless, these filtering goggles reduce the light by about 20 percent. All the studies done in

FIGURE 6-12. Halo and smoke. These two figures each have four rectangles in which the shade of gray changes continuously from one edge to the other. The halo effects on the left and the smoke effects on the right that extend toward the corners of the large squares are illusory (Zavagno).

the laboratory indicate that wearing them does nothing to improve the fine perception of details. But, on the slopes, the hollows and moguls are judged better. By removing the blue, the yellow lenses generate a color signal that helps the interpretation of relief. As in the experiment with noise that becomes louder and louder, we may not be aware of the source of the improvement, and we would be mainly sensitive to an illusory quantitative increase—in one case of sound, and in the other, of light.

If we overlay a page of text with a sheet of highly transparent colored plastic, the readability may improve considerably. The color that provides this benefit differs from one individual to another. In my own case (but I am in a minority) it is yellow or orange. I see better through a yellow or orange transparency, and the printed letters look clearly thicker to me. This effect may be connected with that of irradiation: the black of the letters would normally be trimmed by the white on the page. We would see them as thinner than they really are. The use of a filter, by reducing certain colored components, could thwart the irradiation and restore the letters to their true thickness.

7 SEGREGATIONS, FUSIONS

We are all familiar with those short films made by amateur moviemakers in which the image jumps around and the camera angles follow each other with no logic. Filled with good intentions, the amateur, his eye glued to his camera's viewfinder, moves the optical axis freely following whatever catches his eye: first a head, then part of a wall with a nameplate, then someone's legs. Another person, looking at the same scene without intending to film it, would also have scanned in zigzag fashion, but his "shots" would help to form a stable mental image whereas in the film one view replaces another. The image as a whole that is constructed in our head is an incredible patchwork: the image combines views taken left and right, each view having been itself reconstructed from half-views analyzed separately in the two cerebral hemispheres. The patchwork is pieced together with such care that one sees no seam in it; everything seems as if the image was constituted all at once (Figures 7-1 and Figure 7-2).

A page from a book held in one's hand appears flawless throughout, the letters on it appear printed with the same sharpness, and you have the sense that, given enough time, you could make out all its contents. Kevin O'Regan maintains that this conception is mistaken, that at a given moment we really see only the spot we are looking at, the rest of the image being a largely illusory construct. When one holds a bottle in one's hand, one has the sensation of holding the whole object, a sensation that makes the bottle present beyond the area of contact between the glass and the skin.

In my opinion, what happens is that we assemble small elements into a homogeneous image. The assembly effects are easy to demonstrate. For example, let us write on the front of a rectangular plate the odd-numbered letters of a word, and on the other side, the even-numbered letters. When we spin the plate around its horizontal axis, we see the whole word. Another example: we are seated in the subway when another train passes close by; we are able to see the scene on the other side of this train as something clear and continuous, although we receive only tiny pieces of it through the windows that flash by.

A small experiment that is easy to carry out at home demonstrates the visual assembling of fragments, that of the magician's wand: one inserts a slide in a projector, projects the image in the air, and rapidly moves a white rod across the beam of light. At each instant, a portion of the image appears on the white wand, but one sees, if the rate is fast enough, the entire image formed, suspended in the air thanks to the wand. We are surprised by the image created on the magician's wand. But after all, hearing constantly per-

FIGURE 7-1. The illusion of the pierced hand. Make a tube of small diameter with a roll of paper twenty to twenty-five centimeters long, apply the tube to the right eye while holding it in your left hand, or vice versa, and aim at a reference point a few meters away, keeping both eyes open. "It will seem to you that the right eye, confined to the tube of paper, does not see the object; the left eye alone perceives it, and it appears to distinguish it through an opening in the left hand. The left hand holding the cylinder will seem to be very clearly pierced, and will look like the drawing above." (Gaston Tissandier, *La Nature*, 1881)

FIGURE 7-2. Thickenings. Very slight thickenings in the circles form the message $e = mc^2$. To see it better, give this figure a sweeping rotation without changing its orientation (as when one shakes a glass of water to dissolve a pill in it). You will also see four sectors, two dark and two light, that sweep over the figure in the direction of the movement imposed.

forms the same trick, by making us hear in a unitary way words, sentences, or themes that have come to us in little slices of time.

What chaos there would be in our memory if each heard sentence was stored there, packed together with the interference that had accompanied it: the starting up of a household appliance, the neighbor's toilet flushing, or the car alarm in the street. Perception constantly sorts out, brings the important signals to consciousness, and rejects the rest in an unorganized background. When, during my work I put some favorite music on as a background, I hear the first measures, and then come to the end of the disc with the sense of having listened to nothing, as if the greater part had gone into empty space. But if someone is talking on the other side of a wall, my attention is captured by the voice, which arouses my curiosity and distracts me from the work under way.

In audition, perception also sorts out in order to separate two voices in a conversation, or two musical themes in a polyphony, and this work of segregation sometimes produces illusions. Composers of the baroque era, like Bach and Telemann, exploited such illusions. Several of their scores for a single instrument include passages where low and high notes rapidly alternate. The listener is persuaded that two different instruments are being played, one playing the theme formed by the high notes, the other that of the low notes. The fact that we hear two overlapping musical lines separately is not necessarily an illusion. The paradox is that it is impossible to hear the composite theme formed by the intertwining notes of the two melodies even when these two melodies are unfamiliar and arbitrary. Such a sequence would constitute an "impossible melody." Here a first illusory effect is in the sensation of hearing two sound sources. Another illusory effect results from the fact that we are not able to determine exactly the relative timing of two notes drawn from different themes.

Diana Deutsch did an experiment in which familiar melodies like "Yankee Doodle" were broken up into three levels. A third of the notes was transposed an octave higher, a third of the notes was transposed an octave lower, and never were two consecutive notes kept to the same octave. Subjected to this radical treatment, the melody became unrecognizable. At the most one could guess at it by the rhythmic pattern. But if the listener was informed that he was going to hear "Yankee Doodle," he recognized it right away. All in all, in a first grasping of the melody, what counts first is the "contour," the fact that the notes go up or down in the scale of highs and lows. Listening more closely, where memory comes in, the listener uses chromatic equivalencies, which make a C remain a C, whatever its absolute pitch.

By manipulating a note's environment, we can favor a perception in which the note is located in its natural octave, or favor the one where it is connected with other notes in another octave. In particular, if two notes are separated by exactly a half-octave, the interval can be heard as rising by some people, descending by others. Shepard thus constructed ambiguous musical lines, and Jean-Claude Risset produced one of the most beautiful auditory illusions, that of a continuous sound that rises or descends indefinitely. Risset's illusion used a synthesized sound made up of several notes superimposed in successive octaves. He increased the frequencies of all the notes in parallel, but in this rise toward the high notes, when the highest note went above a certain threshold, he gradually softened it to make it finally inaudible, while a low note was surreptitiously added in order to renew the stock.

A lot of memory and interpretation goes into listening to singing. Indeed, singing is supposed to follow a strict sequence of notes, fixed by the score, where each note has a well-defined pitch, but language is normally expressed through other frequencies: the vowels are more or less high, the "ee," for example, being higher than the "o." So how do we express the note intended by the composer, and make it, according to the need for speech, an "o" or an "ee"? Many singers somewhat master the difficulty by always singing the same vowel, which is not noticed, for we interpret the words instead of listening to them for what they are. Next comes the problem of making oneself heard over the orchestra. The low frequencies go off in all directions; the high ones are more directional and are better-heard by the audience (if the singer projects to it). So it is the high-pitched harmonics that carry best over the orchestra, and the listener nevertheless hears the low-pitched sound, as in the phenomenon of the "missing fundamental" (chapter 2). The differences we perceive between a tenor, a baritone, and a bass are perhaps due less to the pitches of the sounds produced than to certain relations between the frequencies, which we connect to the various ways of using the voice.

Our ability to pick out the signs of language is remarkable. We easily separate two texts written one on the other, when the letters come from different hands, which would be the visual analogue of auditory segregation. We also separate two figures drawn one on the other with the same thickness of line when each figure is meaningful to us. On the other hand, when we join a word to its mirror image so that the two touch, the word becomes hard to read. The letters of the word lose their individuality in favor of a more complex symmetrical unity (Fig. 7–3).

In nature, symmetry is a visual clue of the highest importance. A form with vertical symmetry is almost certainly that of a living animal, including a potential prey or predator, thus a form to which one cannot be indifferent. In most cases, it is of vital interest to detect rapidly the object exhibiting such symmetry in the environment and to react appropriately. The hypothesis,

FIGURE 7-3. Symmetrized letters. The column of forms to the left of the name Descartes contains that of a person often mentioned here. The name is easier to detect when it is horizontal. If it does not pop out, hide its lower half.

DESCARTES

which is made automatically, is that what is symmetrical belongs to an object, that there thus is no need to break it up; on the contrary, it should be considered in its unity, and a meaning attributed to it.

This principle appears in the paradox of the mirror-image letters. It also plays a role in the kaleidoscope, where with some colored discs and two mirrors, patterns of surprising richness are created. Without the mirrors and the symmetry they provide, we would have not patterns, but congregated discs. Symmetry is also exploited in the Rorschach test: a large drop of ink is splashed on a sheet of paper, which is then folded to make symmetrically spread-out shapes. It thus produces evocative forms that psychologists use to assess the mental state of their clients.

I have developed another way of using symmetry to produce evocative images. I use handmade random textures, and I put a band of textures side by side with its symmetrical band or, sometimes, the same symmetrical band on each side of the initial band (Figure 7-4, Plate 10). Faces emerge readily from mottled textures, pseudo-cartoon characters with textures formed of thick scattered lines. Insects and Chinese motifs are also abundantly found. The symmetry is perceptible at once in the immediate neighborhood of the axes; gradually, the organization extends to the edges. As with the kaleidoscope or the letters, the detail of the textural elements is lost in favor of the figure taking form.

When one turns the page and brings the axes of symmetry to the horizontal, the area seen as symmetrical is reduced to insignificance, on both sides of the axis, and one seeks rather to interpret each band separately. In nature, symmetry in relation to a horizontal plane indicates a stretch of water. What is significant—apart from the water itself—is the part located above the stretch of water, which is interpreted without appeal to symmetry. Thus we have two ways of seeing symmetrical figures: when their axis is vertical, we unite the halves on both sides of the axis in order to see a unitary form. When the axis is horizontal, we are interested in this axis, but treat the image like that of an entity split in two.

Figures containing strong contrasts, with regular and close alternations of light and dark, were highly developed by the painters and graphic artists of

FIGURE 7-4. Forms suggested by a vertical symmetry. This image is made up of three equally wide vertical bands side by side. The central band was symmetrized once toward the left and once toward the right. The bands on the left and right are thus identical. The appearance of each of the bands is lost in favor of the patterns that are constructed around the two axes of symmetry and spread toward the edges. When one turns the figure ninety degrees, the symmetrical forms shrink, and one becomes more attentive to the details of the texture, outside the axes of symmetry.

op art. Among the designs that these graphic artists proposed, one started a career among the professionals of perception. The motif had been published by Hajime Ouchi, among hundreds of others, and without any particular note in a Japanese collection, then repeated in a popular American collection for graphic artists. A circular area filled with a checked pattern stretches horizontally and is surrounded by a circular band that is also textured in checks but stretched vertically (Figure 7-5). When the design is moved, with a back-and-forth horizontal movement, the center area seems to dissociate itself from the page and slide onto the pattern surrounding it. As often happens, it is an illusion that one has to learn to see: some people are not susceptible to it at the beginning, then find the right way of moving the image and obtaining the best effect. The illusion is also obtained with a fixed image, provided that one applies light pressure to the edge of the eye repeatedly. I found the same effect in the design of a painting by Reginald Neal in 1964 (Figure 7-6).

FIGURE 7-5. Floatation 1. When this pattern, created by Hajime Ouchi, is moved back and forth horizontally, the central disc dissociates itself from the page and slips onto the surrounding texture.

Pure stripes are generally ineffective, the configuration of Neal's picture being an exception. The checked look—or, more exactly, the existence of elements with clear borders—favors the illusion (Figure 7-7). Everything hangs on the border between the two patterns. The movement may involve a large part of the image. Two portions of the image, which contain different orientations, can move simultaneously in opposite directions. Often, the illusory movement has a unifying aspect: separated areas that contain elements oriented the same way move in parallel.

It may be that this illusion is the signature of a natural process that aims to segregate the environment into significant sets. When we are walking in the street, the scene gradually changes, and the parts nearby change more quickly than the parts in the distance. In general, these changes are not perceived as movements, but one can see movement in a particular case: when a nearby surface interposes itself in front of a part of the more distant land-

FIGURE 7-6. Floatation 2. When this pattern is moved, you can see an area containing one kind of oblique lines (+45 degrees or −45 degrees) float in relation to the areas containing obliques of the other kind. A canvas by Reginald Neal, *Space of Three* (1964), consists of nine of these patterns juxtaposed by symmetry.

FIGURE 7-7. Floatation 3. Pattern created by the author. When the sheet is moved, a central disc containing elements of several different orientations floats in relation to a background also consisting of several orientations.

scape. For example, I am heading toward a cliff parallel to the line of a crest. Visually, this line separates two textures, that of the ground at my feet, and that of the terrain lower down. When I keep my attention on this border as I walk, I see the lower texture move. Another circumstance: if I am moving toward a telephone booth—which partially conceals the front of an apartment building—and if I keep my gaze on its outline, I see the apartment building in the background move. One thus has, when one moves, a certain latitude for assessing the apparent changes in a scene, either as unremarkable modifications, or as apparent movements. The question is then one of understanding why the Ouchi figure promotes the latter, an interpretation of apparent movement of one texture in relation to the other.

As the Mach bands were the signature of an important "neuronal process," that of lateral inhibition, I am led to believe that the Ouchi illusion is the signature of the method used by the brain to calculate movements. Imagine that, to measure the speed of cars on a highway, we placed two pressure-sensitive cables across the road five meters from each other. If the two cables indicated a pressure almost simultaneously, it must be that a vehicle with front and back wheels about five meters apart crossed them, and if a second pressure appeared on the forward cable a tenth of a second later, it must be that the speed of the vehicle was about two hundred kilometers per hour. By placing pairs of cables at various spacings along the road, we could determine the speeds of vehicles of all sizes. But when one wishes to compare the speeds at a particular point, one can place only a single pair of cables, and this latter is well-adapted only to a single length of vehicles. If it is crossed by two vehicles that are going the same speed but are of very different sizes, they are liable to be counted as having different speeds. The Ouchi illusion fits this description well, for the simultaneous presence of differently spaced elements on both sides of a line strongly stimulates it.

8 COMPLETIONS, CREATIONS

In defining *illusion,* dictionaries usually insist on hallucinatory phenomena: to see something that does not exist, to hear a sound that corresponds to nothing. In reality, perception is often led, in its interpretative work, to fill in the gaps, and thus to complete the signals from the outside by internal productions. This is what happens permanently in the filling-in of the blind spot. An area of the retina with a diameter of a half-degree (2.5 millimeters seen at thirty centimeters) does not react to light, for it acts as exit door for the optic nerve. We are not conscious of this hole in our visual field, and we fill it in by extending as best we can the information received around the hole (see Figure 2-1). More generally, when signals received from outside match up with a familiar object, our perception adds to it with details drawn from memory. There is thus a whole progression of effects going from the filling-in of the blind spot to hallucinations and dreams—where the brain produces internal images that impose themselves with the realism of scenes experienced in real life.

The French physicist André-Marie Ampère, who was interested in perception and who in 1805 introduced the word *cognition* in its current sense, gave the example of a French audience member listening to an Italian opera:

> If the words are not pronounced loudly, the listener seated at the back of the hall receives only the impression of the vowels and musical modulations, but does not hear the articulations and consequently does not recognize the words. Let him then open the libretto and follow along with his eyes; he will hear very distinctly the same articulations that a little while ago he was unable to grasp.

A less expensive modern equivalent of this experiment is the foreign film shown in its original language. If you have only a little familiarity with the language in the film, the dialogue is sometimes a confused babble. But when the subtitles clarify (without our realizing it) the general sense of the dialogue, you hear the words with crystal clarity, as if their meaning had been obvious all along.

One develops a taste for a piece of music by listening to it several times. The melody gradually takes form by inscribing itself in our memory, perhaps in the way a scene that is at first sight blurred through poorly adjusted binoculars becomes clear as we bring it into focus. Once the theme is fixed, we listen to it with our memory: the sounds received awaken the theme inscribed in us, and we no longer have the sensation we had when we first heard it.

I am often able to notice the confusion between memory and auditory perception. Listening to a very well-known record in one room of the apartment, I leave humming along with the singer. Entering another room, I am still humming in accompaniment with the song coming faintly through the walls. There comes a moment when I am seized by doubt: am I keeping up with the singer? I don't know whether he is doing the verse or the chorus. I return to the room where I started, and the sense of confusion increases, for the sound becomes more audible but formless. It takes me several seconds for the music to take form and for me to clearly recognize the melody, also extremely well-known, of the next song.

This experience is very instructive. In the beginning, when I move away from the loudspeakers, I receive a signal that is weaker and weaker, and more and more contaminated by new ambient noises, but strong enough for me to reconstruct the known words rather accurately. Then the reconstruction departs from reality. Between what I really hear at the start and what I hear later with a little imagination or what I hear at the end through pure imagination, I am not aware of any qualitative transition. Conversely, on the return trip, although the acoustic signal is increasingly clear, I do not decipher it, not having found the right equivalent in my memory.

In any case, numerous experiments show that you can hear a sound signal as if it were complete even though you have received only fragments of it.

The earliest observations of this kind are related in Radau's *L'Acoustique*, published in 1867:

> A very odd phenomenon is the one that Mr. Willis referred to by the name of *paracousis*. Here is what it consists of. Certain hard-of-hearing persons who usually do not hear faint sounds, suddenly do hear them when they are accompanied by a loud noise. Mr. Willis knew a woman who was always attended by a servant with the job of beating a drum when somebody was talking to her: she then heard very clearly. Another person heard only when bells were ringing. Mr. Holder cites two other similar cases: that of a man who was deaf when one did not beat a nearby bass drum, and that of another person who heard best when he was in a carriage that was jolting over the cobblestones.

The interpretation of this paradoxical phenomenon is that the noise, adding to the words, conveys fragments of the speech above the hearing threshold. It would be on the basis of these fragments that the hard-of-hearing woman would reconstruct the whole of the discourse that she would then have the illusion of hearing continuously.

In a very curious modern variant of this experience, subjects with normal hearing were asked to try to lip-read someone they knew, first when the speech was inaudible, and then when the acoustic speech signal was replaced by a signal without meaning but with the same timbre as the voice of the person being lip-read. In the latter case, the ability to understand the message increased very significantly.

In the laboratory, R. M. Warren did an experiment that has become a classic. He tape-recorded the word *legislature* and erased a syllable. One then heard the two remaining fragments of the word. But when he recorded a very loud noise in the place of the missing syllable, the listeners distinctly heard the whole word and the noise, without locating the one in relation to the other. The brain considers that a sufficiently loud noise, superimposed on a syllable, can totally mask it. If the syllable is not heard, it is still possible that the whole word was pronounced, and the brain exercises

FIGURE 8-1. The gentlemen in evening wear look complete although they are depicted only by white fragments. Their hands, for example, are unconnected with their bodies. Drawing published in *Les Amusements de la science* (Paris: De Savigny, 1905). Bibliothèque Nationale de France, Paris.

the right to complete the signal by forming a hypothesis about the missing part. Similar experiments were done by Giovanni Vicario for musical sound: if in a rising scale, one of the notes was replaced by some white noise twenty decibels louder, one hears a note of an intermediate pitch between the ones that had been retained on both sides of the noise. Although there is in principle a possibility of choice, one always perceives the natural note that is missing from the scale. Other experiments reveal the acoustic equivalent of the phenomenon of transparency, one sound appearing to continue under another.

Artists and scientists are equally fond of the images with "illusory contours." The best-known variant is that of the "Kanizsa triangle," (Figure 8-2) where in the center of the figure one sees a triangle a bit lighter than the background, and whose sides look clearly defined. The outline of the triangle is not drawn. It is only suggested by the circular areas that appear to have a sharp notch cut into them at its border. The other variant is that of Ehrenstein, where the outline of a drawing is suggested by the interruption in a field of dense parallel lines (Figure 8-3). These effects are currently utilized by graphic artists, notably to suggest letters in relief (Figure 8-5), and draftsmen have long known how to suggest a complete form by actually representing only a part of it (Figure 8-1). The illusion is twofold: on the one hand, it rests on the existence of outlines, and on the other, on the difference in brightness between the figure and the background.

Among scientists, there are two opposing conceptions, and the debate has given rise to a multitude of variants intended to determine what was essential for the production of each of the two effects. For one camp, the illusion is cognitive. The interruptions of the lines, the missing sectors of the discs

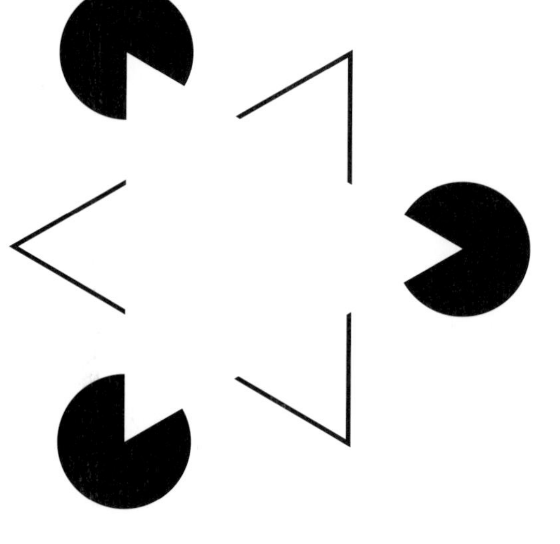

FIGURE 8-2. The Kanizsa triangle. The three indented sectors in the three black discs form the corners of an illusory white triangle. The interrupted lines, which seem to belong to a true triangle located behind the white one, strengthen the illusion but are not indispensable.

suggest that there is a junction between the two surfaces, or the interposition of one surface in front of another. In the Kanizsa example, a triangle would be perceived because a triangle, assumed to be above the drawing, would explain the reason for the very special form and arrangement of the elements that induced the illusion. Against this conception is that of the supporters of a much more elementary, noninterpretive process known in scientific jargon as "low level." With highly technical supporting arguments, they contend that early neuronal mechanisms can form outlines by connecting adjacent inducing elements. The outline is formed by interpolation, as in the exercise where a child must use a pencil to connect a set of dots to produce a figure that is revealed only at the end.

Their strongest argument in favor of "physiological" mechanisms is that in the figures with illusory contours a very faintly contrasting luminous point is easier to detect when it is located on a contour. As the mechanisms for perceiving contrast intervene at an early stage of visual interpretation, this

FIGURE 8-3. Illusory contours à la Ehrenstein. It seems that a draftsman wanted to join the ends of adjacent interrupted lines with as regular a stroke as possible.

result suggests that the illusory contour is constructed even earlier. To this, proponents of the cognitive approach are tempted to reply that the subject's attention is focused on the illusory contours and that this is why he detects more easily the faintly contrasting points. The same debate could be transposed to Vicario's illusion (Figure 8-8): there, where stripes are seen as wider—which is an illusion—they are also seen from a greater distance, which seems to be a physiological criterion.

In my opinion, the illusion of the subjective contours has two ingredients. On the one hand, it reflects essential perceptual mechanisms. Culturally, we have been influenced or misled by the practice of drawing. When one draws an object, one represents its external contour. This contour does not in general correspond to any real line, any distinctive sign that the object bears. The contour is an abstraction that describes the border between what is seen of the object and what is not seen. This border shifts when the object is

FIGURE 8-4. Illusory disc and bands. These forms of very great regularity stand out. A line joining the ends of adjacent lines would have produced much less regular forms.

FIGURE 8-5. Subjective letters. On the lower line, the name GREGORY appears in subjective letters, only suggested by the black forms. These forms act like shadows, and although they delimit the letters only on one side, the letters appear complete. In the example of the upper line, inspired by Kanizsa's *Grammatica del vedere*, the figures at first do not seem to mean anything. But once we have learned to read the reclined letters ELFE, the letters become visible as subjective surfaces (except, in my case, the L, which remains a black outline).

turned or examined from a different angle. In nature, the contour of a zebra is indicated by the break of its stripes. It is thus probably by applying a low-level procedure that perception seeks such cutoff points, that it develops some elements of a possible border. In this, the supporters of the "low level" theory are partly right. However, the candidates for border elements, to be validated, are not combined by a process of interpolation (the pencil that would join the dots). It clearly appears, on the contrary, that the brain produces a plausible surface, sketching borders as best it can, by using a kind of fabrication like that involved with the circles of Figure 8-4 or the letters of Figure 8-5. The strongest example in support of this idea is a variant created by Harris and Gregory (Figure 8-7a).

We are concerned with a pair of images, which makes two subjective surfaces appear. This pair forms a stereogram. When we see one image with the left eye, and the other image with the right eye, we see an illusory surface in relief, with the form of a horse saddle. This surface could neither be interpolated nor calculated point by point, for the two original surfaces are of a uniform black; they contain no mark making it possible to calculate the third dimension. The form of the saddle is thus wholly determined by what

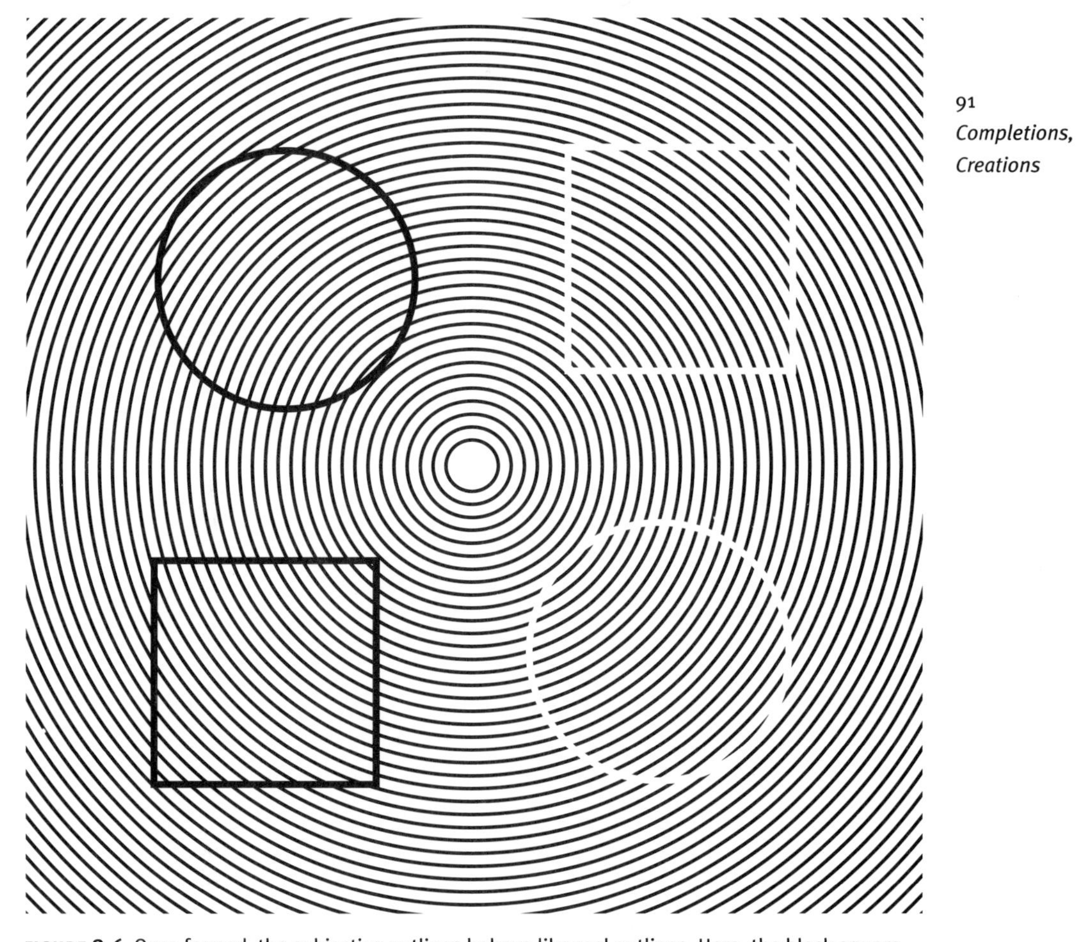

FIGURE 8-6. Once formed, the subjective outlines behave like real outlines. Here, the black square and circle are illusorily bent out of shape because they overlap the concentric circles (Orbison's illusion). The square and the circle with subjective white outlines are illusorily distorted in exactly the same way. Note that the black circle and square are described as real, and the white ones are described as subjective. This distinction may be more cultural than perceptual.

FIGURE 8-7A AND 8-7B. Subjective surfaces in three dimensions. The upper stereogram was created by Harris and Gregory. The combination of the two images produces, in stereoscopic vision, an illusory surface in the form of a horse saddle. The lower stereogram was created by Idesawa and Zhang (1997). It represents a subjective surface in the form of a sphere upon which some cones are set.

happens at the level of the indented discs. In stereo vision the two edges of the notches become the two sides of an angle that projects in the third dimension. The surface perceived has the form that a rectangular sheet of plastic would take whose four corners were inserted into the angles.

Masanori Idesawa has created even more elaborate variants, in which we see two illusory surfaces in space that intersect and we can perceive their intersection, doubly illusory. He has also created subjective surfaces in the third dimension (Figure 8-7b), which are not prolongations of the inducing elements but rest on them as in Ehrenstein's images with subjective contours.

A kind of irrepressible association, called "synesthesia," affects at least one person in every two thousand: those so affected hear "colored vowels." Rimbaud's sonnet of "vowels," which dates from 1871: "black A, white E, red I . . ." is, for these persons, to be taken literally. This phenomenon is six times more frequent in women, is genetically transmitted, and appears in numerous variants. In certain persons the colors are evoked by letters, numbers, words, or even musical sounds. Several composers, including Liszt and Scriabin, were susceptible to it. In others, colors are evoked by tactile or gustatory sensations. In some cases of synesthesia a visual or auditory stimulus triggers a tactile sensation; in others an odor triggers a corporal sensation of movement. The most frequent synesthesia by far is the one in which a language sound (vowel, consonant, syllable, or word), but not its written form, triggers the perception of a color. The first scientific description was made by an Austrian physician, Dr. Nüssbaumer, in 1875, in the *Medical Week of Vienna.* In 1883 Galton conducted an investigation in England and found four subjects who associated the sound "o" with the color white and four oth-

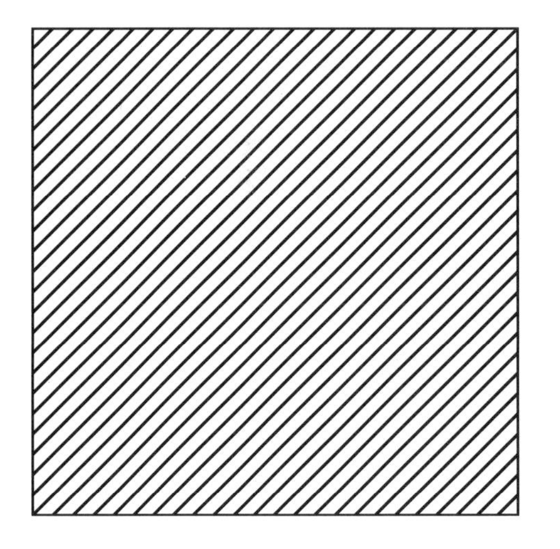

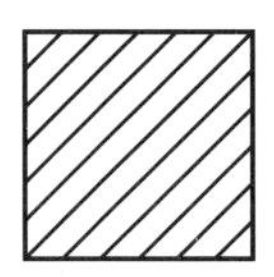

FIGURE 8-8. Vicario's magnifying glass. The lines in the smaller square illusorily look farther apart than those in the larger square. Viewed from afar, there is a distance at which the larger square appears to be a uniform gray, whereas one goes on seeing the hatched aspect of the smaller square. The effect is also observed with a small square inserted in a large one.

ers who associated it with other colors. I assume that the sound "o" in Galton's study must have been the diphthongized English "o," rather different from the French "o." The most serious recent study conducted by Baron-Cohen in Great Britain indicates that there is a very strong tendency there to associate the "o" with white. Otherwise, the most frequent associations were those of "u" with yellow or light brown, and of "i" with white or pale gray.

The colors are always seen "inside the head"; they do not occur in the external scene. In some people, the color does not occupy any particular place, but for others, it has a definite location—for example, a bit above the center of the visual field. The colors arise automatically on hearing the word. The subjects think they have experienced the phenomenon since very early childhood, as far back as their memories go. The fact that such a precise shade of color is associated with this or that word is very reproducible and is encountered again after an interval of some months.

It has been thought that these synesthetic associations develop during the learning of letters, because the letters would have been presented to the child in colored form. A more probable explanation is that when certain areas of the brain begin to take responsibility for language, they do so by overlapping a part of the areas previously devoted to color. The synesthesias would occur when neurons devoted to color, then reallocated to language, would nevertheless have preserved their former links to the areas for processing color.

9 ADAPTATIONS

Perception tends to treat what is permanent as background noise, and to tone it down. Hence our insensitivity to our body odor, forgetting the contact of clothing, or of glasses (which I sometimes look for even though I have them on my nose). The tip of the nose, which is permanently present in the visual field, is eliminated. It reappears once an anomaly occurs. One day, while swimming, I was surprised to see a red spot moving in front of me a few inches from my eyes. In fact, I had slightly injured myself without noticing it, and had a drop of blood on the tip of my nose.

We end up no longer hearing familiar noises, except those conveying particular messages. In the series of little noises in the early morning, one is awakened by the ones that occur shortly before the alarm clock is to go off. Karl Pribram relates that in New York there was an elevated subway above Third Avenue that made a hellish racket. When the line was torn down, numerous phone calls began to arrive at the police station in the middle of the night: people living in apartments along the line had been awakened during the night, in the middle of a peaceful sleep, with a sense that some weird and indescribable event had happened. In fact, they had been bothered by the absence of the familiar noise of the elevated subway.

Certain adjustments of perception are very quick, causing many aftereffects in which a strong sensation in one direction is immediately followed by a sensation in the opposite direction. For example, a person is asked to dip his left hand in a bowl of hot water, and his right hand in a bowl of cold water and then both hands in a bowl of lukewarm water; the water in the lat-

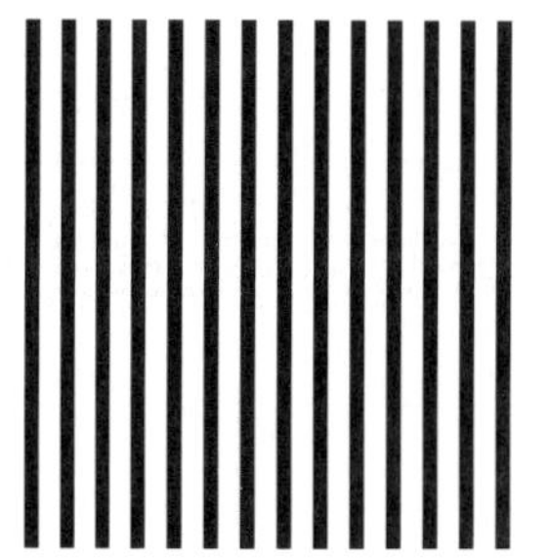

FIGURE 9-1. Adaptation to orientation. Stare for a minute or two at the oblique lines of the figures on the left while hiding those on the right, and then look at the latter: the vertical lines become oblique in a direction opposite to the one of the first image.

ter seems cold to the left hand and hot to the right hand. If the subject is blindfolded, he is convinced his hands are in two bowls.

In the same spirit, museums of science present paradoxical experiments on judging weight. Two cylinders with the same diameter slide on a metallic shaft, the one underneath being longer and hence more sizable than the upper one. The museum-goer is asked to lift the set of them; he thus picks up the cylinder underneath, and makes it slide toward the top, bringing along the other cylinder in the movement. Next he is asked to lift only the upper cylinder; astonishingly, this small cylinder seems heavier than the set of two cylinders he has just

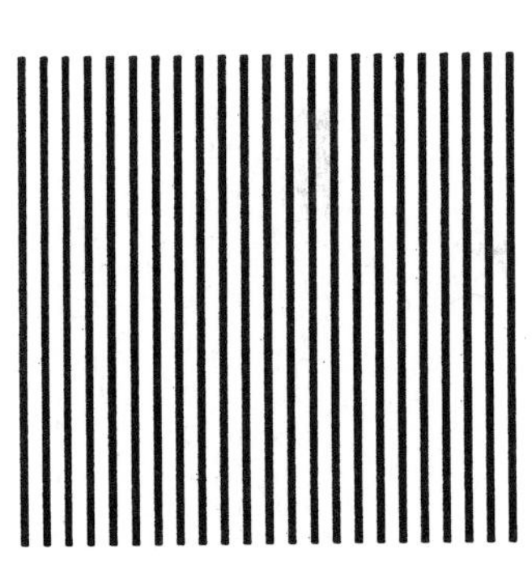

FIGURE 9-2. Adaptation to spatial frequency. Stare for a minute or two at the left half of the figure while hiding the right half, and then look at the right half: the lines of the upper pattern will look closer together than those of the lower one.

lifted. In fact, the cylinders look the same but the smaller one is much denser than the large one. In lifting both cylinders, the subject relates his muscular effort to the overall volume, and this effort is felt as moderate. When he lifts the small cylinder, his muscular effort, in relation to the volume lifted, is more intense, and this results in the illusory sensation of moving a heavier object.

When we look intently at an object of a certain color, then direct our eyes to a white wall, we perceive an illusory afterimage of the object in its complementary color. Here is a nice description of it by Goethe:

One evening, finding myself at an inn, I looked for some time at a well-proportioned servant girl, who had a dazzling white complexion and black hair, and was dressed in a scarlet corselet. She had entered my room, and I stared at her from a certain distance in the half-light. As soon as she left I made out on the white wall in front of me a black face surrounded by a light-colored halo, and the clothing of the clearly drawn figure was a beautiful sea green.

This kind of effect is observed on awakening. During sleep, the eyes have adapted to darkness. In the morning, if the bedroom is bright, you see a glaring light on opening your eyes and you easily form an afterimage. Often, for that matter, the person reporting these images tells how, having looked

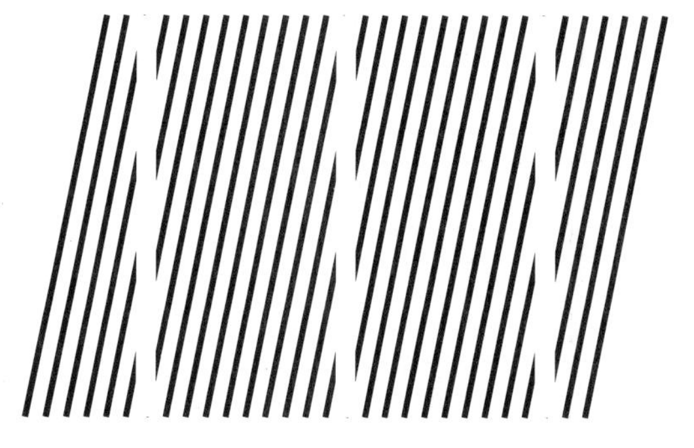

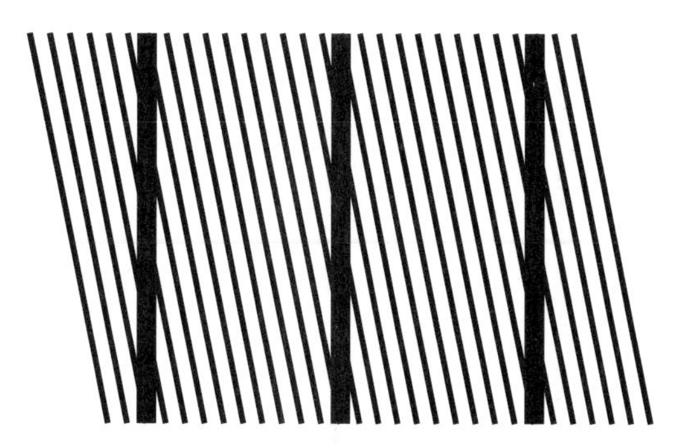

FIGURE 9-3. White or black vertical bands on a background of oblique lines appear to lean slightly in the opposite direction of the oblique lines. Some persons consider that this illusory effect is of the same nature as the adaptation effect of Figure 9-1.

at a window and then at a white wall, he formed the negative image of the window on the wall. Another favorable circumstance is when you look at a television set in a dark room after dozing off for a while. When your eyes leave the screen, you take along a dark spot in the form of the rounded rectangle of the screen. The most effective way to have an afterimage is to use a flashbulb. In a dark room you illuminate an object with the flash of an ordinary camera, and you then form the afterimage of that object with no difficulty.

The afterimage is externalized: it seems to belong to any light-colored surface we look at. It is very probably due to adaptations in the retina. When the afterimage is formed on a sheet of paper, and you move the paper away or toward your eyes, what happens? If you didn't know the answer, you could imagine three possibilities: (1) the sensitized area of the retina remaining the same, nothing happens and one goes on seeing the spot with the same apparent size; (2) because, according to the laws of perspective, the apparent size of the sheet decreases with the distance but the sensitized area of the retina remains the same, you should see the afterimage grow larger, the father away the sheet, and decrease in relative value when you bring the sheet closer; (3) if you think that subjectively the sheet seems to stay a constant size as you move it closer or father away, it could also be that the afterimage remains constant. The right answer is, according to Emmert's law, the second one: you see the afterimage grow larger as you move the sheet away.

If we look at an object with only one eye, its afterimage is apparent whether the eye is then open or closed. Convinced that the light sensations came to us through the eye that remained open, we unhesitatingly attribute the afterimage to it, whereas the afterimage has its source in the eye that saw the object; it is a second-degree illusion, and Newton was taken in by it. There are two important general aspects to this observation: the first is that one can rarely distinguish, in what one sees, between what comes from the left eye and what comes from the right (there are thus people who have lost sight in one eye and notice this only by chance one day when they have deliberately closed only the other eye); the second is that the afterimage from the closed eye is integrated with the normal image of the environment, seen with the other eye, without any sign indicating its different origin.

Some phenomena of adaptation have also been found in audition, notably with the experiments of Georg von Békésy that go back to 1929. If we listen to an intense sound of 800 hertz for two minutes, we will then hear the sounds of this frequency, along with those of neighboring frequencies between 600 and 1000 Hz, as less loud. Moreover, there is a second, more surprising effect that is valid only for the ear to which the initial sound was presented: a second sound, of a different frequency, will be heard with a changed pitch, in the direction of a greater separation from the first sound. A sound of 1,200 Hz will be heard as if it were 1,260 Hz, and a sound of 500 Hz will be heard as if it were 470 Hz. Another effect, described by Zwicker, and the visual analogue of which is not known: if in some "white noise" a band of frequencies is eliminated, and subjects are adapted to this noise, they later hear in its absence a slight ringing at the frequency eliminated.

To create acoustic afterimages requires technical know-how. The equivalent of the white wall would be a "white noise," a noise containing just about all the spectral frequencies at the same intensity. If we listen to a sound with a very characteristic spectral composition, and then to a white noise, the white noise seems to contain the spectrum complementary to that of the first sound. More strikingly, one can construct a first sound that is constituted like the complement of a vowel. After listening to it, one hears the vowel in a white noise that follows it.

The long-term effect, described by Celeste McCollough and which bears her name, has caused a great deal of ink to flow since its announcement in 1965. When one stares for a few minutes at a figure bearing a grid with black and red vertical bands and a grid with black and green horizontal bands (Plate 6), and then gazes at a figure with black and white grids of the same orientation, the white vertical intervals take on a tinge of green and the horizontal ones a tinge of pink. Once experienced, the effect may last for hours. With a training session of two and a half hours, the effect may last for more than a week. The colorings are probably of retinal origin, are not transferable from one eye to the other, and they turn with the head: one thus creates paradoxical images that change color when they are rotated ninety degrees, and also change color when one leans one's head ninety degrees. The Mc-

Collough effect could be related to adaptation to chromatic aberrations. Like poor magnifying glasses or poor lenses that add colored fringed edges to black and white print, the optics system of the eye breaks down the light and does not reconstitue it precisely on the retina. The blue is diverted toward the periphery in relation to the red. Colored fringed edges form where there is a sharply defined border between light and shadow—hence the utility of an adaptive mechanism blocking these fringed edges. Someone wearing glasses whose lenses create colored edges adapts to them and no longer sees them after two to three weeks.

At the end of the nineteenth century George Stratton forced himself to wear an apparatus that inverted images: the device consisted of two short tubes fitted with lenses that formed inverted images, reversing high and low as well as left and right. The tubes were fixed to the head in such a way that only light transmitted by the lenses entered the eyes. For the few days the experiment lasted, Stratton restricted himself to seeing everything through his reversing device. A wrong account of this experiment is often given, reporting that after a few days or weeks, not only did Stratton quickly get used to the situation, but also that the image of the external world received through the instrument had righted itself once again and that he then saw the world the way he had seen it before the experiment. In fact, for Stratton as for those who have replicated this kind of experiment, the adaptation has not seemed to involve the righting of the image.

In the beginning, Stratton made the following observations: the whole visual field looked inverted, what was up was seen as down, what was on the left was seen on the right; when he turned his head or body, the whole visual field seemed to turn in the same direction but much faster (which is explained by simple geometrical considerations). There was a dissociation between sound and sight: a source of noise was heard in one direction and seen in another. There was also a dissociation of touch and sight: an object in contact with the body was felt in one place and seen in another. In particular, Stratton felt his own hands or feet as elsewhere than where he saw them. His memory could not guide his movements inside his house, and he had to proceed by trial and error.

Gradually, by forcing himself to use vision to guide his actions, certain incongruities subsided. For example, if he tapped on the arm of his armchair with a pencil point, the sound seemed actually to come from the point. The sensations of the body fit into the visible world: in walking, he saw his feet hit the ground, which appeared in the upper part of the visual field, and the sensation became normal. Space was reorganized and he was once again able to plan his movements inside a room or from one room to another in his apartment.

At the end of the experiment, when Stratton removed the lenses (but kept wearing the cylinders to preserve the same visual field), the world seemed strange but objects were not inverted. The main aftereffect was that the scene turned when he turned his head or body, but this problem disappeared from the second day on.

In Austria, in a variant of Stratton's experiment, Ivo Kohler and his colleague Erismann had their subjects wear prisms that reversed left and right without inverting high and low. After a while, the subjects could move without banging into things and could even ski or ride a motorcycle. At the start of the adaptation, Kohler reported that furniture seemed correctly distrib-

FIGURE 9-4. Aftereffect of movement. When one makes Plateau's spiral (1850) turn clockwise—for example, on the turntable of a record player—the whorls appear to spread outward. If you stare at the spiral for a few seconds and then stop the movement, you perceive the reverse contraction movement of the whorls; the spiral still appears to have a constant size.

uted in a room, that on the road cars indeed seemed to be driven on the right, but that the inscriptions on facades or license plates on cars were still seen in their reversed form. Toward the end of the adaptation, the subjects had a sense of complete normality even for inscriptions. But doubts remain; according to Charles Harris, the subjects never restored the inscriptions left to right but became so used to reading them in reverse that they paid no attention to the direction in which they were written.

A still-gentler version of the experiment consists of wearing prisms that shift the visual image sideways. In the very beginning, when the subject is asked to point to a target or to throw darts at it, he systematically aims to one side, but then adapts to the displacement. When he takes the prisms off, he makes the reverse error but soon gets back to normal. The prisms have other effects that are harder to deal with, however. First, they break up the light and produce colored fringes plainly visible in the border areas where there is a strong contrast in luminosity. The adaptation to these fringed edges takes several weeks; after reflecting on this phenomenon, McCollough thought of constructing the experiment described previously. In addition, the prisms caused distortions in the image; when the subject turned his head to the left or right, he experienced an "accordion effect" and when he inclined his head, he saw the scene swing to and fro like a rocking chair. Several weeks were also needed to get rid of the accordion and rocking-chair effects.

A common point in all these experiments is that the subject makes mistakes primarily in the assessment of the position of his body; for example, he believes he is looking straight ahead, though his head is turned in the direction that offsets the effect of lateral displacement caused by the prisms. Harris shows quite convincingly that the perceptual reorganization occurring during the adaptation is primarily a relearning of the positions of the hands, feet, and head in relation to the visual signals. Contrary to all the old ideas about "the education of sight by touch," these adaptation experiments suggest that sight strongly dominates the sense of touch, and that the main component in adaptation is a reorganization of bodily sensations so as to adjust to the new visual order.

Finally, here is a real-life experience recounted by one of my colleagues. As a child, he had an undiagnosed astigmatism for horizontals. When it was detected—he was then twenty-five—and he was made to wear glasses, he at first saw forms compress. The dishes he knew to be round and had seen as round now looked elliptical, and his mother's face, which he had seen as long, became round. A few days later round objects had gone back to being round, with or without glasses, and his mother's face remained round. He could never see it again as he previously saw it. This experiment is doubly paradoxical. If there is a compensatory mechanism for the effect of compression produced by the glasses, how does it happen that the forms are the same with or without glasses? I am led to see in this phenomenon the implementation of an important principle of perception, according to which perception manages, as much as possible, to provide conclusions that are independent of the way in which the signal was processed, which incidentally makes the identification of this process difficult.

IMAGE 1. Picture from the "Enigma" series by Isia Leviant. After approximately ten seconds, you see flows of matter spinning in the rings.

IMAGE 2. The stream going uphill. This little stream in Yosemite Park is flowing toward the background. In the photo, it seems to fall toward the front. The illusion is probably due to the constriction created by the two light-colored boulders in the foreground. The stream gets wider as it gets away from the observer, and the widening is wrongly attributed to an upward slope of the surface. (See also Figure 11-4.)

IMAGE 3. *A Peepshow with Views of the Interior of a Dutch House* by Samuel van Hoogstraten, painted around 1756, a masterpiece of trompe l'oeil. Through a peephole in a panel closing the front of the box, which was here removed, the viewer saw himself inside an apartment room opening onto other rooms, the different volumes perfectly well-arranged. In fact, there is only a single volume, and certain objects in it are painted at the junction of two oblique panels, or on three panels spaced in depth. The National Gallery, London.

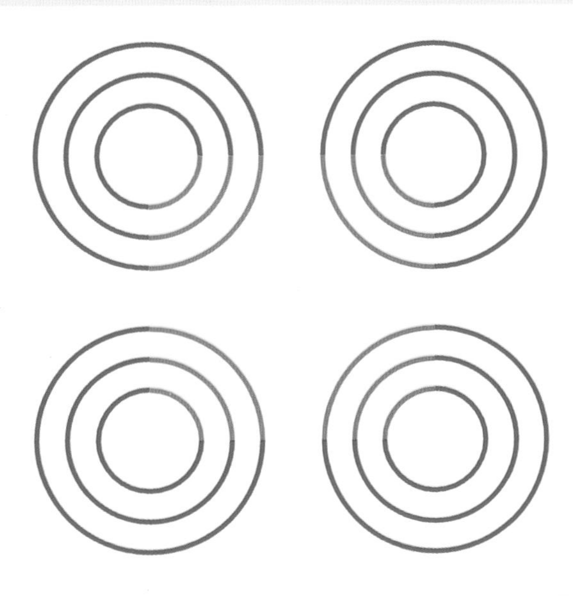

IMAGE 4. Dario Varin's illusion, or "neon effect." The arcs of green and red circles create the illusion of a square surface of pink.

IMAGE 5. Contrast effect. This figure is formed of simply juxtaposed rectangles of uniform color. If you cover the separation between two adjacent bands with one finger, you see the shades become equal on both sides. You will also notice an effect of shading-off: the bands seem to grow lighter on the side of their border with the darker adjacent band.

IMAGE 6. McCollough effect. Look attentively for at least five minutes at the patterns with vertical red and horizontal green bands while hiding the black-and-white pattern opposite. Look next at the black-and-white pattern in dim lighting. You will see pale green horizontal stripes and pink vertical ones appear in it. When you bend your head ninety degrees, the illusory colors keep their orientation, horizontal or vertical in relation to your head, and thus change location on the pattern.

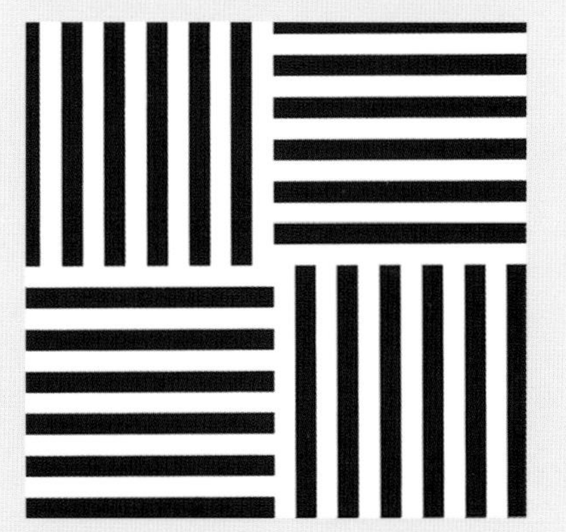

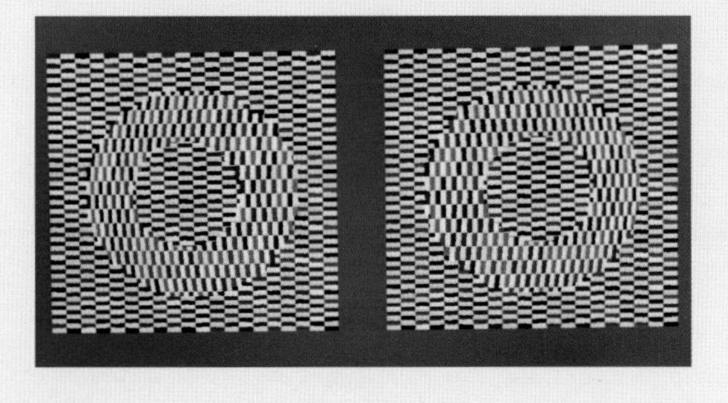

IMAGE 7. Floating cone. When merged in stereoscopic vision, this pair of images produces a surface in relief in the form of a lampshade that seems to float above the background when the page is given slight horizontal movements.

IMAGE 8. The pointed squares of Isia Leviant. This image was made according to the principle of the "Inferno" series of pictures and silk-screen prints of Isia Leviant. When it is seen from far enough away (three or four meters), the squares lose their regularity and appear to have pointed corners.

IMAGE 9. Three-dimensionality with a single eye. Observe the protuberance of the seaweed through a narrow tube, or your hand cupped into a cylinder in front of your eye. You will see the protuberance rise up and take on a three-dimensionality similar to stereoscopic relief.

IMAGE 10. Suggestions. This image contains a central band formed by juxtaposing rug fragments and two lateral bands that are the mirror image of the first one. Patterns take form along the vertical symmetry axes, and are more elaborated the longer they are observed. They shrink in favor of the original pattern when one turns the page ninety degrees.

10 CONSTANCIES

Even at high speeds, some drivers enjoy turning their heads to admire a precipice by the roadside or to tell a joke to a passenger in the back seat. The feat performed by perception in this case is to supply the driver with a stable image of the road despite the head movements. Let us consider some less critical circumstances. Peacefully seated in a room, I look around: the scene appears to move and to change slightly. In general, I am oblivious to these changes in appearance. Normally, things appear to move very little when I nod my head. If, however, while closing one eye I briefly press the lower lid of my open eye, I clearly see objects move and return to their initial position. The explanation accepted by specialists is that under normal conditions, eye movements are planned by the brain to explore the environment; knowing the position to which it directed the eye, the brain foresees the way in which the image may be changed and makes the necessary corrections so that the image appears stable. When you press a finger on the eye, the eye goes through unplanned movements, and the brain does not make any correction; under these conditions it is the environment that seems to move.

Through perception we want to identify places, objects, and persons, so we have to store their permanent characteristics in our memory and to recognize them despite changes of appearance. Perception provides a standardized representation, endowed with *constancy,* with all the effects of context corrected for, and thus comparable to the representations held in memory. The constancy effects are found everywhere: in colors, brightness, sizes, forms, sound intensities. The one notable exception is the estimation of speed. According to our philosophical predilections, we can classify a constancy effect—or a lack of constancy—as an illusion. Constancy being what perception provides naturally, analytical effort is required to understand that it follows from a mental process, and imaginative effort is required to construct the experiment that puts this constancy in check or diverts it to paradoxical effects.

We know that color, as it is perceived, is corrected for variations in the ambient light. At dusk, the light takes on a predominantly red shade that shows up in photographs but has little effect on the tonalities perceived. Ibn el-Haytham was, in 1040, the first to mention the phenomenon of color constancy. La Hire, in 1694, evoked it as follows:

> There is nothing that the eye gets used to faster than the change of color. One can very easily do the experiment to show it by looking through a piece of glass slightly tinted with green or some other color & by hiding objects that can be seen other than through the glass. Shortly, you will no longer notice that all the objects are tinted green or some other color. You will notice it even less if you put the glass in front of your eyes after closing them for a long time, and before opening them.

He also explained that by candlelight a blue object turns green, but we go on seeing it as blue, and he proposed an experiment to convince oneself of this. Go into a room lit by daylight, and observe through a keyhole the objects in an adjoining room lit by a candle, and compare the colors: then, you will see that these objects "will appear tinted with a reddish yellow compared with those that are lit by the sun & can be seen at the same time."

Because of constancy effects, the real difficulty is to recover raw sensation. This is why Helmholtz recommended assessing the real tonalities of a landscape by looking at it with one's head between one's legs. Commonly too, painters turn their pictures upside down, the better to appraise the color harmonies in them.

The moon reflects light received from the sun, which is yellow, and sends it back yellower still. Nevertheless, we generally see the moon as white. When I question children about the color of the moon or the stars, I am surprised to hear them answer fairly often that they see them as yellow. We freely admit that the moon looks yellow in hazy weather, but this color is rarely ascribed to stars. Here, however, is a description of the night sky, by Fontenelle, in which the colors are distributed in an unusual way:

> The Moon had risen perhaps an hour earlier and its rays, which came to us only through the branches of trees, made a pleasant mixture of a strong bright *white* with all that green that looked black. There was not a cloud that hid or obscured the littlest star, they were all of a pure and dazzling *gold*, and which was even enhanced by the blue background to which they were attached.

Size constancy has been discussed many times since Ptolemy. The size of the image that an object makes on the retina must vary inversely with the distance. Still, at eight feet (three meters) a face looks the same size as at two feet (one meter) and not a fourth the size (in diameter). Size constancy works for nearby objects in a visually rich environment. If I move my thumb closer to or away from my eyes, it stays the same apparent size. But if I go through the same movements while observing it against a uniform background, through a darkened tube (or a hand forming a tube), I see its size change. My thumb shrinks as it recedes, as perspective would predict.

Seeing objects as constant requires a compensatory mechanism correcting their apparent size according to their distance. What happens when the distance is not really known or not known precisely enough? Gregory ventured the hypothesis that the brain uses, as a substitute, indications of perspective: certain configurations of lines would be typical of projecting or recessed objects or of scenes staggered in depth. According to him, in the case of a geometric figure shown on a flat sheet of paper, certain configurations would also be treated there as signs of perspective and would involve, by the illegitimate application of the mechanism of size constancy, an increase or decrease of size, and thus it is that most of the geometric illusions would arise. (See, for example, Figures 10-1 to 10-3.)

Maffei and Fiorentini made an interesting observation: size constancy applies to the dimension of screens. A television screen seen close-up remains small, but a movie screen seen from the back of a theater looks much larger, even when it covers a smaller visual field. The result is that the persons on the television screen are definitely small and those on the movie screen, exaggeratedly large.

The indications of perspective can be manipulated to create illusions. The most common form is the kind of image in which we are shown two elements of equal size in a scene containing strong indications of perspective. The one that is interpreted as farthest away is seen as larger than the other (Figure 10-4). For the demonstration to be complete would require that the effect be canceled when one turns the page around, since in this case the perspective should no longer hold. I leave it to the reader to judge.

Size constancy applies poorly in the vertical direction. The persons on an airplane-passenger bridge look small when seen from below, although they are not very high up. According to Buffon, we have learned to estimate sizes and distances in the horizontal direction, but we have no equivalent experience in the vertical direction. For this reason:

> when we find ourselves on top of a tall tower, we judge the people and animals below as much smaller than we would judge them at an equal dis-

FIGURE 10-1. Scale corrections and inverted perspective. These stacks of parallelepipeds, made of Lego blocks of uniform size, seem to get wider toward the base, in both the photo and in reality. But in the photo the real width at the base is 8 percent less than at the top. This is probably one of the numerous variants of the chevron illusion. This variant was discovered by Gregory, who discussed it as follows in Gregory and Gombrich, 1972: "A new illusion. This striped rectangular parallelepiped is made from the Lego construction game for children. Toward the bottom it seems to expand away from the observer—contrary to perspective. . . . Possibly the many crossed contours lead to corrections of scale appropriate to a longer object."

FIGURE 10-2. The spacing between the two structures in the foreground looks greater than the width of the first structure on the left, whereas it is just slightly less. The illusion may be due to the interpretation "in depth" of this spacing.

tance that was horizontal—that is, in the usual direction. It is the same with a rooster or a finial that one sees on top of a church steeple; these objects look much smaller than we would judge them indeed to be if we saw them in the usual direction and at the same distance horizontally at which we see them vertically.

The anomalies of "upward constancy" pose a problem for sculptors. The sculptor who wants to erect a statue on a pedestal introduces compensations: he enlarges the upper parts of the statue in relation to the lower parts so that, seen from below, the statue looks well-proportioned. The practice is unlike what we know of "horizontal" perception, in which we accept seeing things as they are. More precisely: the sculptor finds it normal to enlarge a rider in relation to his horse in order to compensate for the decrease in size

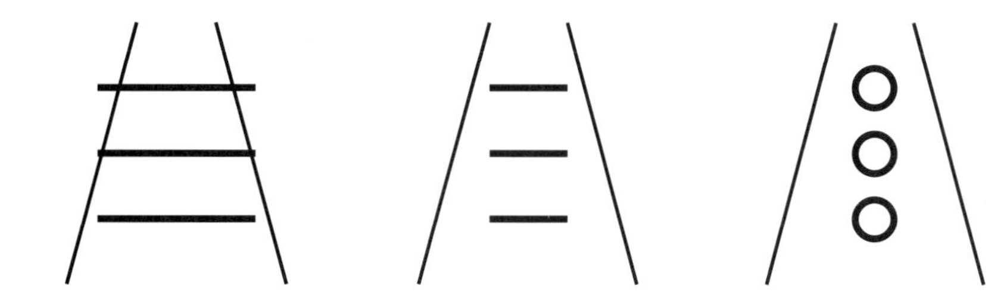

FIGURE 10-3. Geometric illusions and perspective. The two white rectangles on the photograph are the same size, but the upper one looks larger, which is consistent with an interpretation in perspective. Some writers have proposed that geometric illusions appear in configurations with strong perspective clues. Thus, the drawings of the left or center (above), where the cross lines are of the same length, make us think of a road or of train rails in perspective. The upper cross line looks longer than the lower cross line, but the illusion is clearly weaker than on the photo. Note that in both cases, the illusions persist when you turn the page around. A drawing similar to the one on the left, illustrating an article on the relation between illusions and perspective, appeared in *La Nature* (1893, second half-year, pp. 315–318), signed by Charles Edouard Guillaume. The variant in the center is known as Ponzo's illusion (1928).

owing to the distance in height. But in a statue made to be seen head-on (for example, a sphinx at the entrance to a temple), it does not occur to any sculptor to make the back bigger relative to the front. The practice of vertical compensation lacks consensus. Vergnaud treats it harshly in his manual on perspective:

> The colossal statues placed on too-high buildings from too close up are . . . a concession made to the thoughtlessness of the public, to its habit of looking at buildings from too close up and thereafter inevitably raising their eyes to see the statues that dominate them; a habit that moreover the public is forced to adopt by clogging up the areas around these build-

FIGURE 10-4. The skiers of Lycée Buffon High School. The figures of the skiers, set in the upper gallery of the Lycée Buffon in Paris, are all the same size. But the one farthest away looks much taller than the others, because of the strong perspective cues.

ings in such a way as to make it impossible to place oneself properly to consider the whole from its true point of view.

The most striking of visual constancies is no doubt that of forms. If a figure merely gets closer or farther away in a frontal plane, all the parts keep the same relations and the figure keeps the same aspect. A circle always looks like the same circle whether nearby or far away. But a three-dimensional object changes aspect with the distance, as perspective dictates: the frontal elements diminish inversely with the distance, whereas the thicknesses decrease inversely with the square of the distance. (See Figure 3-9.) The practiced eye can recognize on a photograph whether the face was photographed from up close (in general, with a short focal distance) or from far away (with a long focal distance). It is often said that short focal lengths make facial features stand out whereas long focal lengths soften them. In fact, it is the distance of the camera shot that matters and not the focal length. The less-practiced eye becomes aware of these effects under extreme conditions: when, to produce a comical effect in the movies, a face is filmed from extremely close up and then the nose or chin assumes a disproportionate size.

We also compensate for changes in appearance caused by inclination (Figure 10-5). A round plate is seen as round even when viewed from an angle in which, objectively, it should appear elliptical. Form constancy doesn't perform as well for rotations of a figure in a plane. A face turned over is poorly interpreted and, in general, forms are in this case significantly modified. We recall the square-diamond illusion (see Figure 3-10). Turning practically anything deforms it: there are various illusions of spiral contraction and expansion (see Figure 9-4), ellipses that turn become slack, and regular polygons that turn behind small openings also seem to change form or size.

Perception cares little about durations and even less about speeds. In this matter, it is precise only to help us react to danger: to protect us from a stone thrown toward the face, to foresee the moment of impact in a jump to assure a good landing, to estimate the margin of safety when we are crossing a road where cars pass. But perception is not reliable for telling us at what speed we

FIGURE 10-5. Shape constancy. When one looks at this photograph absent-mindedly, one sees identical tiles with square motifs that cover the walls and bottom of the basin as well as the central block. But, because of the strong perspective, the squares are more or less distorted parallelograms, and one has to make an effort in order not to see them as squares. For example, on the right-hand face of the central block, the squares are compressed toward the bottom, and on the left-hand side of the same block the squares are narrow, almost rectangular parallelograms.

are moving nor to estimate the speed of objects that are receding or moving transversely.

Sensations of speed come primarily from the body. In the trains of old, the passenger would open the window and, leaning outside, receive the bracing slap of the wind in his face. In the moving sarcophaguses of today, the train windows are hermetically sealed and all physical contact with the outdoors is broken off. We are informed that the train moves along at 180 miles (three hundred kilometers) an hour or that the airplane flies at 600 miles (one thousand kilometers) an hour, but no visual sensation enables us to confirm that the figures have not been doubled.

Our culture has a fanatical interest in speed as measured by the stopwatch. We consider that a champion who has swum 100 meters in 47.95 seconds—that is, at a speed of 7.51 kilometers an hour—has performed a tremendous feat, outdistancing by a tenth of a second the next swimmer, who has swum the distance in only 7.49 kilometers an hour. This difference makes sense for the sportsman, who knows what it has cost him to earn this tenth of a second, but the reference is once again corporal and not visual.

We know that with distance the apparent size of objects must diminish, but we are fooled by the fact that the apparent speed also decreases in the same proportions. Thus, Flaubert writes in *A Sentimental Education*, describing a horse race in a hippodrome:

> From afar, their speed did not seem excessive; at the other end of the Champ de Mars they even seemed to slow down and to advance only by a kind of sliding movement. . . . But returning quite quickly, they grew ever larger; their passing stopped the wind, the ground trembled, the pebbles went flying. . . .

If I come close to a fly that is moving on a wall, I do not see it speed up its movements. Speed constancy must exist at very short distances; at intermediate distances, size constancy would apply but that of speeds would no longer be involved. Watching a television broadcast of an auto race or tennis match, the viewer does not have the sensation of speed that he would have if he observed the event life-size. In tennis, if the spectator is seated in the bleachers some twenty to thirty meters from the court, which has an average length of twenty-four meters, he sees the balls fly by at an angular speed ten times greater than the one seen by a viewer sitting three meters from a television set, where the image of the court would be about thirty centimeters.

We also find constancy effects in audition. Probably the most curious effect is that we are able to recognize that people are speaking loudly, or that music is blasting, even when the sound is muffled. In a concert hall, the sound of voices or instruments comes to us in a straight line and also indirectly by bouncing off walls and other objects. Our acoustical perception removes echoes for us, but these echoes reappear when we are listening to recordings. In general, when the sound interacts with an obstacle, it is reflected with no change of frequency and reaches the listener with a slight lag compared with sound that has traveled in a straight line. Human audition is very sensitive to frequencies, a constant property of sound, and markedly less sensitive to slight time lags, which are highly dependent on the environment. Helmholtz observed this "phase insensitivity," thanks to which, for example, a note from a violin sounds exactly the same whether the bow produced it by being drawn in one direction or in the opposite direction. In the laboratory, when we break a sound down into its various harmonics, we note that the timbre depends on their relative intensities, but that it is not much influenced by their phase lags.

There exists a very strong constancy, equivalent to shape constancy in the visual domain, which is melodic invariance: a melody is still recognizable after transposition (but there are some exceptions).

Professional musicians have often complained that pitch tends to climb with time—that is, established practice has required them to tune their instruments to a standard A that has gone higher and higher. According to Leipp, this is an illusion; with age time is less and less filled, the "psychological second" becomes longer. "Now, an elderly musician," says Leipp, "with good musical memory, has in memory a standard of temporal reference, which he stored away when he was young, at a time when his 'psychological second' was shorter." In support of his explanation, Leipp adds a personal observation that is difficult to get around: after measuring the pitch of sounds during opera performances, he noted that during moments of great dramatic intensity, singers and instrumentalists play higher than was provided for in the score. He reasoned that in these moments of tension their "psychological second" is shortened and the performer has the feeling of producing sounds that are too low; he thus gradually raises the pitch of the sound he produces. Still, music historians remain convinced that the standard tuning pitch has really evolved: the A3 was 404 Hz under Louis XIV, 423 Hz under Napoleon I, and 450 Hz under Napoleon III. Today, the standard A oscillates around 440 Hz, and the inflation seems to be under control.

REFERENCE POINTS

Both our acquaintances and familiar objects appear to us to have the same size inside or outside the house. But there are some objects that one always sees indoors and other objects that never come into the house, and we have a lot of trouble linking indoor measurements to their outdoor measurements. The distance from one wall to the one opposite it in a room that is twelve feet (four meters) long seems substantial, and I have trouble imagining crossing it from one wall to the other in one jump. Outside, at a long-jump pit, the twelve feet (four meters) seem trifling. Without being particularly gifted, the average adult can jump fifteen feet (five meters) and the record jump is over twenty-nine feet (nearly ten meters), the equivalent of two adjoining rooms. When I think of the bed on which I lie down full length and which seems to be fairly large, its length six feet (two meters), I find it hard to realize that it would take two such beds placed end to end to go from the hood to the trunk of a midsize car (Figure 11-1).

The architectural space we live in is dominated by right angles. The rooms are parallelepipeds, and the floors and ceilings are perpendicular to the walls. We tend to assume that everything is this way until there is proof to the contrary, and this leads us to make mistakes in unusual settings that do not respect these constraints. With this idea, Ames constructed an illusion room often reproduced in science museums. This room has trapezoidal walls and windows calculated in such a way that, seen from a particular point—a hole punched in one of the walls—they look rectangular; we think that the walls and ceiling are at right angles, as in an ordinary room. A child standing in the angle where the ceiling is lowest

FIGURE 11-1. Interior/exterior. Perception adapts to the dimensions of the space we move in. An object looks larger inside a room than it does outdoors. Thus, the car shown in the top photo is fourteen feet eight inches (4 meters 46 centimeters), and the bed at the bottom of the page is six feet eight inches (1 meter 94 centimeters) long. The car is thus more than twice as long as the bed. The bed is shown in the center photo at the same scale as the car.

FIGURE 11-2. The Ames room. In this room the windows and walls have trapezoidal shapes, the floor rises to the back, and the ceiling descends. The proportions were worked out so that from a particular viewpoint this room has the same look as a "normal" (parallelepiped) room in perspective. We favor this mistaken impression and wrongly see the person on the right as much taller than the one on the left. Photograph by Philippe Plailly/Eurélios from the Cité des Sciences et de l'Industrie.

looks like a giant compared with an adult standing where the ceiling is the highest (Figure 11-2).

The sky looks slightly concave to us; it is as if it extended far away, parallel to the ground. There would thus be, in our visual space, less space in height than in width. Some people have attempted to use this layout to explain the illusion that makes us see the moon as much larger when it is close to the horizon than when it is high in the sky. The phenomenon in question is indeed perceptual and not a question of the propagation of light through different types of atmospheres. The moon's diameter measured at every height above the horizon varies very little; it is close to half a degree. Beginning in antiquity, the moon-size illusion has given rise to many theories. If we think that the moon is part of a flattened celestial vault, it would have to be less distant vertically than it is horizontally. That is why, at an equal apparent diameter, it would seem larger when it is low on the horizon. All in all, the sky would act, for the moon, like a gigantic Ames room. Another idea often advanced is that when the moon is high, no object comes between the moon and us; but when it is low on the horizon, we judge its distance in relation to terrestrial reference marks that are themselves judged very far away.

For my part, I put forward a factor that to my knowledge has never been appealed to: as regards size constancy, "upward constancy" plays much less of a role than constancy in depth. The moon would be treated like any familiar object or being, thus with a strong correction for size in depth and a lesser correction in height.

The moon illusion is more striking when the sky is a little hazy. This suggests that the effect of "aerial perspective," according to which the farther away an object is, the less contrasting it is, would play a major part in the illusion. The fact that an object looks weakly contrasting thus suggests that it is far away, from which follows, holding size constant, the tendency to see it as larger. The illusion is possibly related to the one known to hunters keeping watch in the morning fog. They sometimes see a bird of great wingspread loom up, but it proves to be a small crow. In this situation, where the hunter is looking up, he does not have a terrestrial reference point to calibrate distances. The size of the bird is then judged primarily by its contrast with the sky; holding size constant, the fog diminishes the contrast, and the bird looks much larger. The same reasoning explains certain hallucinations known to car drivers: often drivers relate that, driving at night, they have seen a huge animal cross the road rapidly a few yards in front of them. The driver does not have time to gauge his distance from the animal, especially as this distance quickly changes; what's more, the driver is misled by the unusual lighting conditions. Having overestimated the distance, he credits the animal with enormous size.

Some people are victims of similar illusions even under conditions in which they have all the time they need to size up a scene. Around 1905 Harvey Carr conducted an investigation with some 350 students, and found fifty-eight of them who said they had experienced illusions of distance. An often-described illusion resembled the one that occurs when one is on the point of fainting, where everything seen recedes and gets smaller while remaining in focus. But the illusion of distancing occurred many times during their lives while they were lucid and in good health. Among some of them, there was an alternation, the scene appearing in turn either close-up or far away. For others, the change was felt as a single overall movement.

In a few, more unusual cases the scene remained stable and only one person approached or receded. For example, one female student saw the preacher in church move farther away, beyond the wall, and remain there for a time, the illusion occurring only at the church, whose appearance did not otherwise change. Another student experienced this illusion in the street. Suddenly someone nearby to whom she was speaking began moving slowly away toward the end of the street—which was not long and ran into a cross street—then slowly came back to his original position. The movement in both directions was perceived distinctly. The "perspective" aspect of the scene was strongly felt—that is, the size of the houses seemed to vary inversely with their distance, and the other person's size seemed to diminish when he moved away and grew larger on his coming back. In several cases described by Carr, the sensation of visual distancing was accompanied by a sensation of auditory distancing: the victim of the illusion had to lean forward and prick up his ears to understand what the person moving away was saying to him.

Finally, of the 350 students, Carr found five able to produce the phenomenon at will. One female student could voluntarily bring objects closer or move them away, even in stereoscopic vision, without this involving any change in the environment. But when she let her attention wander, the scene could move forward or go back as a whole. When the student produced the phenomenon intentionally, the object receding grew in size and became less distinct. In this case it seemed that the object's movement was strongly tied to changes in focusing, but not to changes in the eyes' convergence.

We estimate the vertical dimension using the organs of equilibrium while allowing ourselves to be influenced by visual stimuli that are sometimes mis-

FIGURE 11-3. Sloping pool. This pool in the park in Sceaux, near Paris, does not look horizontal. It seems to form an inclined plane that, for some people, slopes from right to left, and for others, from the front to the back. The illusion is equally strong in real life.

leading, notably where flat-looking inclines in the road are familiar to the thighs of cyclists and to car drivers.

De Gramont gave this rare example:

> In a forest of coastal pines in the Landes, for example, the trees, which have grown in the offshore wind, grow markedly parallel, but inclined toward the interior. We assign an imaginary vertical so well to the set of the trunks that if a house looms up among the trees, the edges of its walls no longer look straight up, but inclined in the opposite direction of the general leaning of the pines.

Similarly, I am susceptible to illusions of water surfaces that, looked at attentively, no longer look horizontal (Figure 11-3, Plate 2).

There is a place, on a road near Banyuls, France, where motorists are tempted to stop to admire the landscape. At this spot, the road is on a hillside and goes into a hairpin bend. The driver who stops and gets out of his car without setting the parking brake is surprised to see his car begin to go forward all by itself on the road that, however, appears to be going uphill. Actually, the road is going down, but various features of the terrain contribute to an illusion of an upward slope. Near the stopping point, where the drivers carelessly park their cars, is a leaning telephone pole that suggests a false vertical, which is consistent with the false horizontals given by rows of vines. Another factor, less easy to spot, may be the principal cause of the illusion. Imagine a very flat road, with parallel sides. Its apparent width diminishes with distance. When the road is downhill, this apparent width diminishes more quickly, and the two sides seem to meet, yielding a plunging effect; when the road goes uphill, its apparent width diminishes less quickly than if the road had been horizontal (Figure 11-4). Conversely, an apparent width at eye level, greater than the expected width, could suggest a road going uphill. If the road really does get wider, which may happen at the approach of a bend, and we think the road is straight, we may have a sensation of upward slope. The road near Banyuls actually did widen as it approached the bend.

One of the first illusions noted in antiquity was the one that in cloudy weather makes us see the moon move as it encounters clouds. The clouds are propelled by the wind, and in daylight their movement is clearly perceived in relation to landmarks on the ground. At night, however, it seems that the bank of clouds serves as a reference mark and what we perceive is the relative motion of the moon in relation to the clouds, which is interpreted as an absolute motion. An amusing variant of this conversion of relative into absolute motion is the pigeon's-head illusion. When we look at a pigeon walking in a straight line on the ground, we see its head characteristically bobbing back and forth. This is an illusion. Actually, the head goes staight ahead with small horizontal movements, then stops, and the feet catch up with it. What we see is a regular movement of the body plus the relative movement of the head compared with the body, which is interpreted as an absolute movement. There are many illusions in the assessment of mo-

FIGURE 11-4. Uphill and downhill. Curves delineated by the sides of the road are automatically interpreted as uphill slopes and downhill slopes. The very rapid convergence of the sides from the start of the road indicates a downward slope; the much weaker convergence in the distance indicates an upward slope. Reciprocally, if a road really widens or shrinks, this may help give an illusory sensation of uphill or downhill slope. (See also Plate 2.)

tions owing to problems of detection, and they are generally not demonstrable on the printed page. I know of two exceptions, one discovered by Isia Leviant, which he calls the illusion of the revolving cones (Figure 11-5), the other one I recently proposed: the hula-hoop illusion, which until now has apparently not been recognized (Figure 11-6).

On an airplane flight, we can see the clouds migrate in various ways as we fix our attention on the plane's wings, the clouds, or the earth below. Rudolph Arnheim noted that when a plane flies above a layer of scattered clouds through which we see the ground below, the clouds may well be standing still, but even so we see them passing over the ground because, in perspective, they move faster than the land does. But if we try to perceive the clouds and the ground as connected parts, the clouds appear to remain fixed and the system as a whole looks stable while the plane flies over it. Even more complicated cases of perceptual reorganization have been described in which two layers of clouds exchange their depths.

FIGURE 11-5. The revolving cones of Isia Leviant. When the page is given a slow up-and-down vertical motion, the two striped surfaces revolve in opposite directions around the axes of cones. This kind of illusion probably results from the absence of reference marks on the stripes, making it hard to assess their movement.

I have made the following observations in a train going over flat country as it drew near a town. The countryside was rather barren, but there were a few landmarks, among them houses at various distances. Very close, the poles next to the rails seemed to be coming toward me, thus to be moving in a direction opposite from the train. At an intermediary distance (about a quarter of a mile [four hundred meters]) the houses looked fixed, like most of the landscape. But the very distant houses seemed to be moving *in the same direction as the train,* though more slowly. B. Bourdon made the same observation. He also noted that sometimes the vertical lines glimpsed outside the train seem slanted. This illusion, he said, occurs on the curves. In these places:

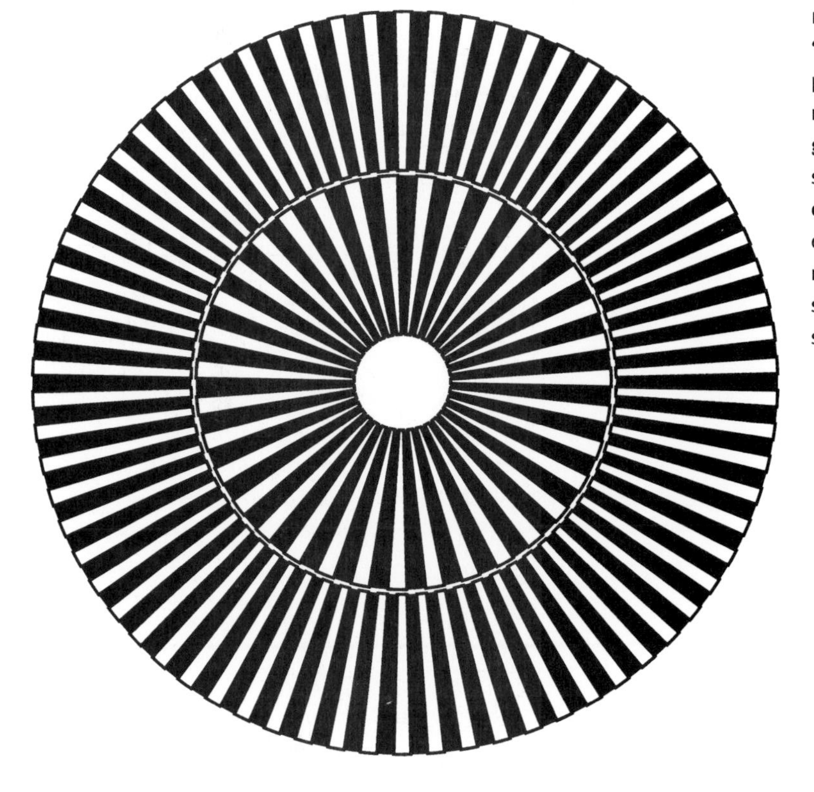

FIGURE 11-6. The vagabond circle, or "hula-hoop" illusion. When the page is given a wide, quick circular movement (the kind of movement given a cup to rinse it or to dissolve sugar in it) without changing its orientation, we see the intermediate circle move along with this movement, like a hoop being spun slowly around the hips. We also see subjective colors near the center.

one of the rails, in order to counteract the centrifugal force, is higher than the other; consequently, the coach we are sitting in and we ourselves as well are tilted a bit to one side. Now we are not aware of this tilt or we take it to be less than it really is; thinking our body still vertical or nearly so, we experience the illusion that the external objects, which really are vertical, are leaning.

I am always surprised by the sensations associated with being passed in a car: when, for example, I am driving sixty miles (one hundred kilometers) an hour and am passed by a vehicle going a bit faster (say, seventy-some miles [120 kilometers] an hour), I first have the sensation of a presence on my left side; for a time that seems normal, then once the vehicle has passed me, I have the feeling that the distance between us grows very quickly and in a few seconds the vehicle is practically at infinity, at the limit of my field of vision. I have the same sensation at sea, when I am passed by a motorboat. Its speed cannot be very great and yet it seems to me that in a few seconds the boat that has just overtaken me has reached a place on the coast that it will take me an hour to reach.

The explanation for the phenomenon could be that, as long as the car is close by, we estimate its distance correctly and it appears to have a more or less constant size. But beyond some fifty yards, we stop applying the correction for size, see it shrink, and have the sensation of a rapid distancing toward infinity. Another explanation, which does not exclude the first one, is to imagine that in this kind of situation we estimate the car's distance according to the direction of the line of sight: if you must lower your eyes to see its rear wheels, the car is close by; if you must raise them up to the horizon, the car is far away. By assuming that the eyes of the driver are also some four-and-a-half feet (1½ meters) off the ground, between five and thirty feet (five to ten meters) the line of sight changes 8.2 degrees, but between ten meters and infinity, it changes no more than 8.5 degrees.

When we are traveling on a road lined with poles of the same height—for example, streetlights—and we look at the nearest one, it appears immobile and of constant height, whereas the next pole, seen in peripheral vision, seems to get smaller and move away though we are actually getting closer to it. When we look at poles lined up at right angles to the road, we see their direction revolve. Mach described the sensations of modification of visual space when he was on a train:

> Supposing that I am going in the direction of the train's movement, the whole space that is to my left, for reasons that are easy to understand, revolves around a very distant axis and in a clockwise direction; the space to my right does the same, but in the opposite direction. Only when I *resist* following objects with my gaze does the sensation of forward movement occur.

A related phenomenon easily observable when we are sitting forward in a vehicle driven with the headlights on at night through a snowstorm is that the snowflakes, though they are falling vertically, all appear to shoot out from a center ahead of us and to come straight at the vehicle. When we stare fixedly straight ahead, the lateral snowflakes even seem to be driven by a slight upward movement.

12 ARBITRATIONS

Information supplied by the various senses is not always in agreement. Where there is disagreement, it is necessary to arbitrate, either by compromise or by giving priority to one sense. The philosophical tradition embraced the primacy of touch and thought that, in the first years of life, sight was educated by touch. In the laboratory, however, we know how to set up situations in which a person feels an object (a cube, for example, or a straight rod) and sees another of a slightly different form (a parallelepiped, a curved rod). Contrary to the philosophers' intuition, we believe what we scc, not what we touch.

For playwrights and novelists, vision, even when mistaken, wins out over the other senses. How many plays are there in which the heroine puts on a man's clothes and conceals her hair, which is enough for her to be taken for a man, without her voice giving her away? How many bawdy tales are there in which a man slips in darkness into the bed of another man's wife without her noticing the substitution, by smell, touch, or even the man's amorous behavior?

Vision also enters into the interpretation of auditory signals; we understand what a person is saying better when we are looking at him. McGurk and MacDonald prepared videotapes in which the image is of a person pronouncing one syllable, whereas another syllable is recorded on the sound track. The viewer-listener is not aware of the contradiction; he hears distinctly a syllable other than the one recorded and that is phonetically closer to the one pronounced on the image. Thus, when the image says "ga-ga" and the sound says "ba-ba," we hear "da-da" with our eyes open and "ba-ba" when our eyes are closed. The localization of the sound source is also influenced by vision: when we listen to a person speaking or singing in front of a microphone, we have the impression that the sound is coming from his mouth, not from the loudspeaker. In operatic choruses, young women of great beauty are sometimes placed in the front who simply mime the singing while the actual women singers are in back. The audience is completely taken in.

In principle, thanks to stereophonic listening, we can assign a direction to the sound source, except for a small ambiguity: the fact that our ears are fixed does not allow us to distinguish the front from the back. Von Békésy recounted that one day a politician had come to see him, very disturbed because when he was listening to some music with a stereophonic headset, he heard the orchestra as playing behind him and not in front of him. His wife had recommended that he see a psychoanalyst but instead he thought of taking his radio to a repairman. Von Békésy proposed a little exercise, at the end of which the patient was quickly cured: he had him sit across from some loudspeakers, then asked him to put his hands next to his ears and to pivot them to pick up in turn the sound coming from in front of him and the sound coming from behind him. In this way the politician could localize at will the orchestra in front of or behind him, and then he learned to make the mental reversal without using his hands.

The two ears do not play a symmetric role in auditory localization, as Diana Deutsch has shown in a very curious experiment. She delivered to the subject's right ear a series of sounds alternating in pitch—high, low, high, low, and so on—and she delivered the same sequence of sounds to the other ear, but with a time lag, so that when the high-pitched sound was received in the left ear, the low-pitched sound was received in the right one, and vice versa. So the subject always received a low-pitched sound and a high-pitched sound at the same time, but the sounds alternated from side to side. What the subject perceived was very different: he heard alternately a high-pitched sound in the right ear and a low-pitched sound in the left ear. Deutsch's clever interpretation of the experiment is that the brain treats the sound as if it were a single, though rich, sound coming from a single source. To localize it, the brain relies on the high-pitched part of the signal, which is logical, for high-pitched sounds are more directional than low-pitched ones. The subject thus perceives a sound that seems to come alternately from the left and the right. As for the nature of the sound perceived (higher or lower pitch), the brain would utilize preferentially the information coming from the right ear. When this ear receives the low-pitched sound, the subject has the illusion of a low-pitched sound coming from the left because the high-pitched sound, on which the localization is based, is then to the left.

Ventriloquists fool us through sight and hearing. As their lip movements are slight or hidden, we readily accept that the sound comes from somewhere else, especially if they are manipulating a dummy whose mouth movements they control. Ventriloquists also work with their voice quality, which they make hoarse, muffled, or resonant to give the impression that sounds have come from somewhere else, the direction always being conveyed by skilled miming.

A point to remember is that perception provides us with the results of its analyses without really telling us by what means (auditory or visual, for example) it reached them. A curious example, cited by Warren, is that of tactile sensations, whose origins would be auditory, experienced by blind people. He says that blind people avoid certain obstacles by detecting sound echoes reflected by their surfaces: the blind person is not aware of the acoustic nature of the signal and perceives it instead as "pressure waves" on the skin, which become stronger closer to the obstacle and "threaten" to become unpleasant if he continues approaching the obstacle.

I would like to explain here geometric illusions in relation to arbitration problems, though that is not at all where they are usually treated. Geometric illusions are small sketches of very simple figures with a small number of lines, in which certain aspects of shape or size are perceived incorrectly. The earliest illusions were discovered by Fick in 1851, Oppel in 1855 (Figure 12-8), Zöllner in 1860 (Figure 12-1)—who mentioned that of Poggendorff (Figure 12-4)—then by Delboeuf, Wundt, Helmholtz, and Müller-Lyer. These are the ones nearly always cited as examples and most-studied by scientists. These illusions have given rise to innumerable variants and more are proposed every year, but it becomes harder and harder to know whether an illusion is genuinely new or merely a combination of effects already recognized.

When we examine an illusion without paying attention to its many variants, it is easy to propose an ad hoc explanation of it. But most of the "obvious" explanations proposed for them by scientists or philosophers can be quickly eliminated.

One persistent explanation is that sizes are assessed by eye movements. Thus, in the classic Müller-Lyer illusion, the eye would go through a wider sweep on the figure with outward-opening angles (the lower figure in Figure 12-5a) than on the figure with inward-pointing arrowheads, resulting in an overestimation of the length of the axis of the former. The Müller-Lyer illusion, like many others, persists in the absence of eye movements, which is demonstrated in at least two ways: one, by presenting the figures only long enough for a flashbulb to flash, which does not give the subject time to explore the image; and, second, by creating a "stabilized" image that remains invariant on the retina when the eye moves. The illusion holds up under both conditions.

Most of the explanations make the angles of the illusion play an all-important role. These are easily refuted when we consider various versions of the Müller-Lyer illusion: the angles are not at all necessary. (For example, Figure 12-5b.)

We also exclude the explanations invoking phenomena occurring at the level of the retina. To prove it, a stereoscopic pair is constructed presenting the illusory figure in some disguise. The figure giving rise to the illusion is not physically formed on the retina, but emerges only at the end of a process of stereoscopic interpretation. Most of the illusions hold up under these conditions. Consequently, they arise at a stage of visual interpretation occurring after the one where images coming from the two eyes are confronted.

Before stating my views on the origin of geometric illusions, I would like to point out some of their general properties. First, people are not equally susceptible to them. An illusion very strongly experienced by one person is not seen by another and vice versa. So we cannot rank the "strength" of illusions on a single scale. Illusions testify to our extreme sensitivity to the shapes and dimensions of figures. When we measure the assessment errors really made, we see that they amount to very little—on the average, an error of 3 percent on the orientation of a line, or 5 percent on the ratio between two sizes.

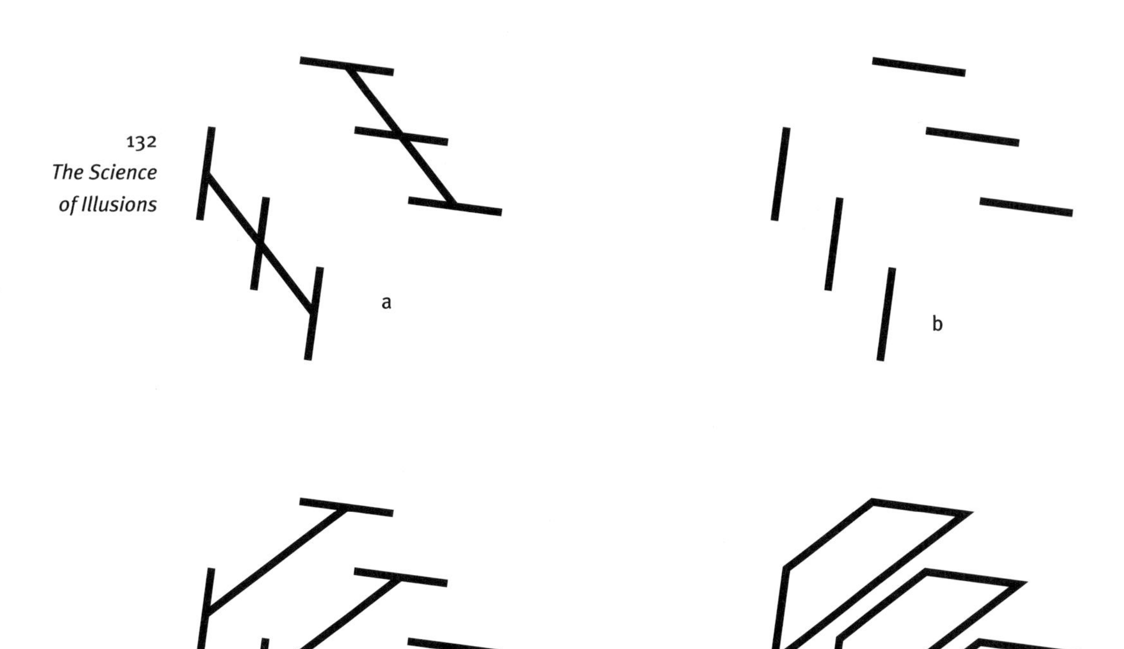

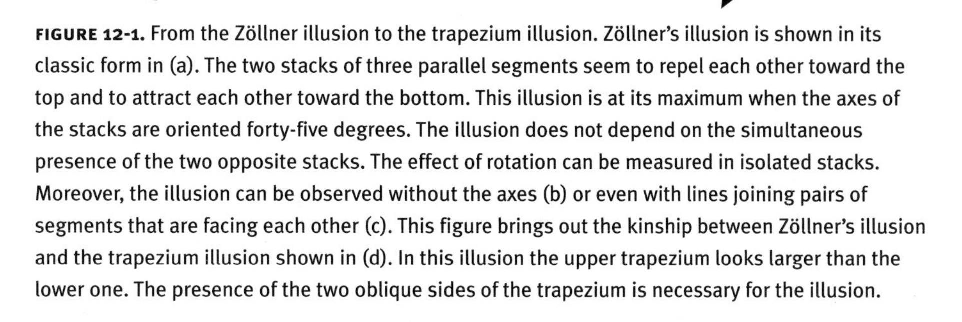

FIGURE 12-1. From the Zöllner illusion to the trapezium illusion. Zöllner's illusion is shown in its classic form in (a). The two stacks of three parallel segments seem to repel each other toward the top and to attract each other toward the bottom. This illusion is at its maximum when the axes of the stacks are oriented forty-five degrees. The illusion does not depend on the simultaneous presence of the two opposite stacks. The effect of rotation can be measured in isolated stacks. Moreover, the illusion can be observed without the axes (b) or even with lines joining pairs of segments that are facing each other (c). This figure brings out the kinship between Zöllner's illusion and the trapezium illusion shown in (d). In this illusion the upper trapezium looks larger than the lower one. The presence of the two oblique sides of the trapezium is necessary for the illusion.

Falling victim to an illusion does not mean that we are not reliable, nor that we have trouble determining proportions and making a decision. For example, I am very prone to the orientation illusions of the Zöllner kind (Figure 12-1), and make mistakes 50 percent above average. But I make

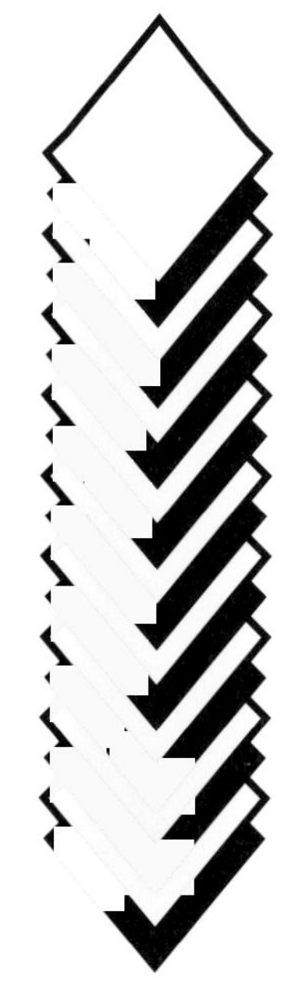

FIGURE 12-2. Variants of the Zöllner illusion. On the left we find the chevron illusion. The stacked-up quadrilaterals are all the same size, but the stack seems to shrink from the bottom to the top. On the right, the "moon's crescents illusion" published by De Savigny in 1905 is particularly revealing. In the illusion, the lower crescent looks larger than the upper one. The explanation of the trapezoid illusion (Figure 12-1d), that the upper trapezoid "encompasses" the lower one, predicts the reverse of what we observe with the configuration of the moon's crescents. On the other hand, the interpretation that invokes Zöllner's effect on the points of the crescents predicts the illusion in the right direction.

these mistakes in a reproducible way and with a feeling of great ease; I have no hesitancy about the way to arrange the blocks so they look parallel to me. I am much more ill at ease when making metric adjustments—that is, setting two segments to the same apparent length, as in the horizontal-vertical illusion. But in these cases I make practically no mistakes.

Illusions have an optional character; an illusion may be manifest in normal vision and disappear in stereoscopic vision. Moreover, if a subject is asked to hold out his hand as though to grasp between his thumb and index finger a figure containing an illusion, he spreads his finger correctly. An area of the brain—concerned with stereoscopic vision in the first case, in manual prehension in the second—thus has available correct information about the sizes of figures, whereas the area that brings the shapes to consciousness introduces distortions in them.

Moreover, our vision of things is fairly coherent, and making illusions contradict one another is difficult. You might think it easy to construct a figure with three segments, a, b, and c, where a looks larger than b, b larger than c, and c larger than a; I do not know of any examples of this kind. More exactly, we cannot make a figure's metric aspects (measurement of sizes) conflict, but we can make a judgment of size contradict a judgment of orientation, which Morinaga's paradox demonstrates. (See Figure 13-1.)

Another important point is that we are very often unable to say *what* the illusion consists of. Take Poggendorff's illusion in which two lined-up seg-

FIGURE 12-3. Illusion. In these three identical superimposed photographs, the central hemisphere looks smaller than the upper one. Courtesy of Cités de la Science et de l'Industrie, Paris.

ments look out of line (Figure 12-4). Is the loss of alignment caused by a translational shift of the two segments or to a rotation? In the first case, we can imagine that the two long parallel lines tend to attract each other, bringing the segments closer together and losing their alignment. We can also imagine a shearing effect, each segment being vertically attracted to the nearest horizontal edge. In the second case, we can invoke a perceptual tendency to see the acute angles as more open than they really are: there would thus be a rotation of the segments around the point where they end on the large lines. We can also invoke a rotation of each segment around its center which brings it closer to the horizontal direction, and so forth. For a few years now I have been conducting systematic work to clarify the nature of the displacements and to substitute for the subjective description of the illusion a precise statement concerning its content. I think that a correct and precise description of the content of illusions will enable us to bring to light their kinship links, to classify them objectively, and finally to help us make progress in our understanding of their causes.

In my opinion, everything that concerns orientations, alignments, and angles has its source in corrective devices to compensate for very real distortions caused by perspective. Perspective modifies angles, makes parallel lines converge, and crushes grids. But in order for the shapes to keep a constant appearance, one must apply to the images distortions complementary to those of perspective. Unfortunately, this idea does not enable us to understand at present how the illusion changes with orientation. Rotating a figure 180 degrees should change things completely, according to a purely perspectival interpretation. This is rarely the case, however—or else these illusions send us back to a far distant time when our ancestors willingly spent part of their lives hanging upside-down from a branch of a tree.

As regards illusions of size, I think I can describe them using just a few particularly simple rules. Specialists have employed two terms to describe the effects of luminance: contrast, the tendency to exaggerate the separation between two juxtaposed shades of gray, and assimilation, the tendency to bring them closer. The same vocabulary is used to describe geometric illu-

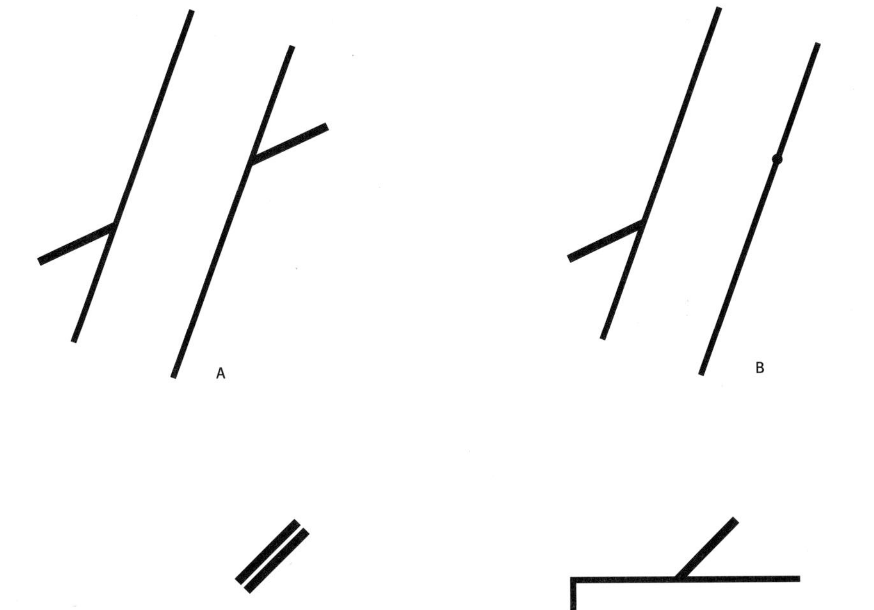

FIGURE 12-4. Poggendorff's illusion. The illusion is shown in its classic form in (A). The two thick segments that end at the parallel lines are aligned, but if we extend the lower one in thought, it seems to end on the right-hand line a bit below its junction with the other segment. In fact, we can already measure an illusion in (B) where it is a matter of aligning a segment with a point, or in (C) where we tend to see the left-hand segment aligned with the lower right-hand segment, whereas it is aligned with the higher segment. In (B) it seems that the error of aiming follows from an error about the angle between the segment and the long line, according to the principle that makes us judge acute angles as more open than they really are. The illusion measured in (A) is mainly the sum of the effects measured in (B) and in (C). Illusion (D) is a corner variant of Poggendorff's illusion discovered by Green. The two segments are collinear but appear to have different orientations.

sions, but in my opinion it is a case of misuse. So I prefer dropping these two terms and starting from scratch.

Imagine a very simple figure containing two segments, of lengths a and b, with b larger than a. My first principle is that the fundamental effect behind illusions of size is the tendency to perceive the b/a ratio as larger than it really is. There would be an exaggeration of the contrast between b and a that generally goes unnoticed but sometimes takes a paradoxical turn when a third size comes into play. In the Müller-Lyer illusion, we often find two segments with the same length (x) but that appear to be of unequal lengths. What has happened? Think about Figure 12-5b. We have two figures to compare, the upper one containing two long segments of length b flanking a segment of medium length x, and the lower one containing two short segments of length a flanking the medium segment of length x. If there is a tendency to exaggerate b strongly in relation to a, but to keep certain proportions in the figures nevertheless, the segment of length x flanked by segments of length b will be somewhat stretched, and the segment of length x flanked by segments of length a will be somewhat shrunken. This description takes account of all the known variants of the Müller-Lyer illusion.

Until now, the most effective concept for describing the illusion was that of assimilation: it consisted of saying that the middle segments, of length x, were

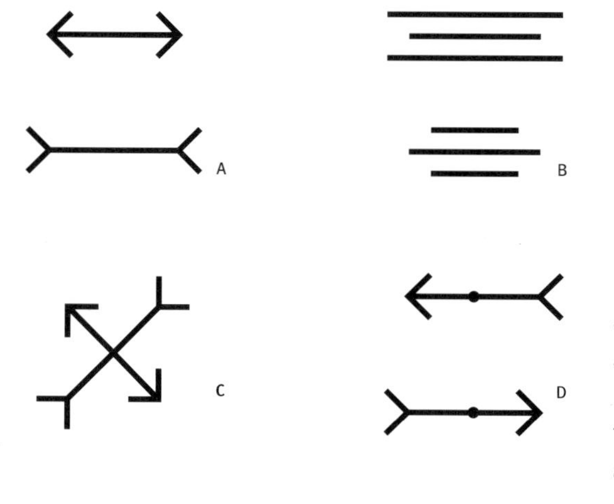

FIGURE 12-5. The Müller-Lyer illusion, probably the most famous of the geometric illusions. The classic form is shown in (A). The two horizontal segments are of equal length, but the upper one looks shorter than the lower one. The illusion is often attributed to the geometry of the "arrowheads" at the ends of the segments, inward pointing in one case, outward pointing in the other. But these V-forms do not play a role in the illusion. They may be replaced by squares or circles, or any appendage to the segments may even be eliminated, as in (B). What counts is the presence of longer or shorter extensions on the sides of the line segments. Illusion (C): a crosswise variant in which the illusion is weakened. Illusion (D): the Judd variant. The points, which are in the middle of the horizontal segments, appear displaced toward the end that bears the V-form.

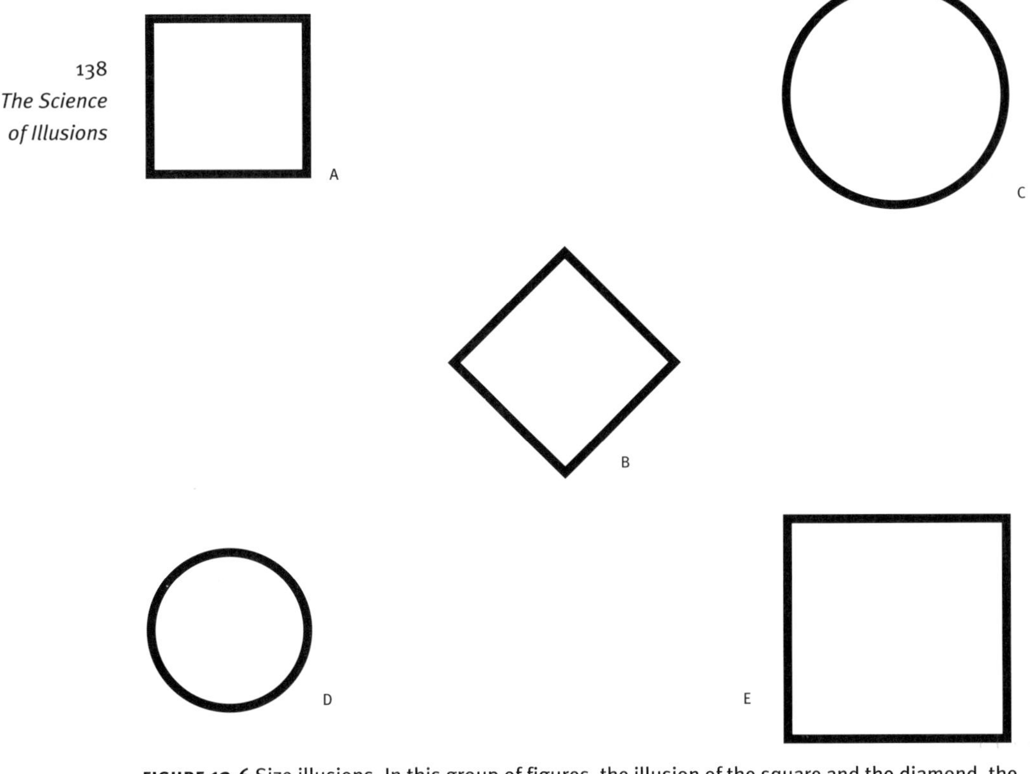

FIGURE 12-6 Size illusions. In this group of figures, the illusion of the square and the diamond, the square in (B), here called "diamond," looks larger than the square in (A), but they are equal. The illusion is based on the fact that the large stretches are overestimated relative to small ones. Thus, the side of the square in (E) or the diameter of the circle in (C) looks larger than the diagonal of the diamond in (B) that is, however, equal to them. Secondarily, we observe that the diameters of the circles (C) and (D) look respectively smaller than the sides of the squares (E) and (A), probably because the squares (E) and (A) contain pairs of points (for example, the ends of a diagonal) that are farther apart than the diametrically opposite points on the comparable circles (C) and (D).

pulled in the direction of the neighboring segments. To this way of seeing things, the tendency for the outer segments of lengths a and b had to be for the *reduction* of the differences between them. According to my hypothesis, the driving force behind the illusion, on the contrary, lies in the tendency to

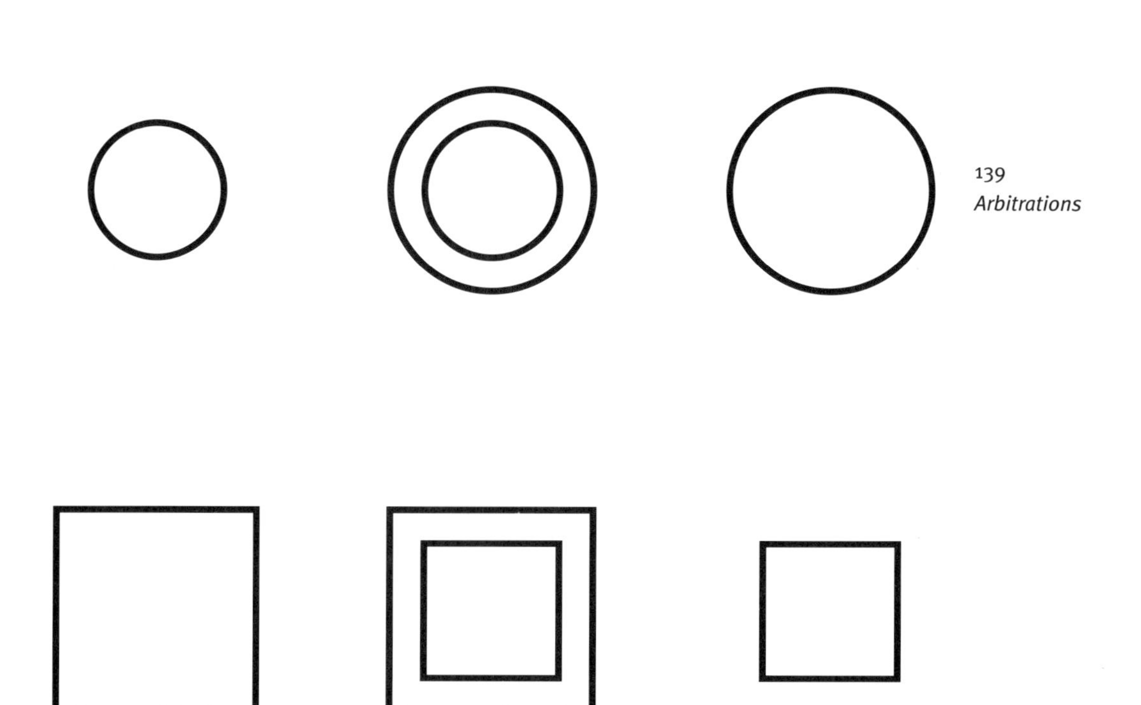

FIGURE 12-7. Delboeuf's illusions. The two concentric circles in the center are equal respectively to the circle on the right and the circle on the left. But the isolated circle on the left looks smaller than the internal circle in the center, and the isolated circle on the right looks larger than the external circle in the center. An analogous phenomenon is noted with the squares.

heighten the contrast between the extreme segments. This allows me to understand (or even to construct at will) a whole series of effects in other figures. Take the example of the square and the diamond (Figure 12-6). The small square looks smaller than the diamond, in accordance with the classic illusion. But now I consider a second, larger square whose side is equal to the diagonal of the diamond (or that of the small square). This time, the illusion is in favor of the square, whose side consequently looks larger than the diagonal of the small square. The same effect is obtained when the squares are replaced by circles (Figure 12-7). This shows that the exaggeration effect of the contrast is stronger than the illusion present in the square-diamond comparison. Once understood, this effect is easily met with in other geometric illusions.

The second tendency I see at work in geometric illusions is a tendency toward *normalization.* It is a matter of normalizing the figures so as to put right, on a large scale, the cumulative errors made on a smaller scale. This normalization is expressed in two forms. When a figure is finely subdivided, because each subdivision is underestimated, there must be cumulative error. Thus a compensation device is introduced by which the figures subdivided into many parts are perceptually enlarged. There thus would be, in these figures, an illusion caused by overcorrection. (For example, see Figure 12-8.) This is the well-known effect in dressmaking involving stripes: horizontal stripes lengthen the figure; vertical stripes thicken it.

Besides, if we consider the total space on a page taken up by a figure (itself composed of several figures), the viewer would normalize the figures, taking into account the overall space occupied. This is what would happen in the illusions of the Titchener type (Figure 12-10). Here again I propose

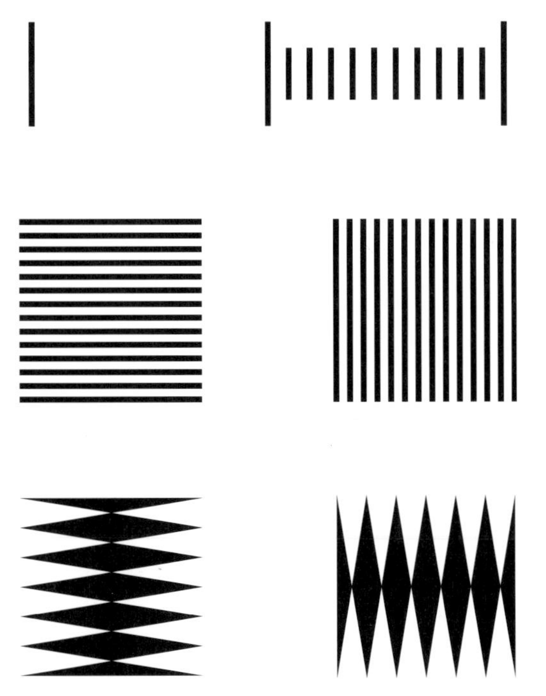

FIGURE 12-8. Expansion of subdivided surfaces. Above, the Oppel illusion is shown in its classic form. The subdivided part of the figure on the right looks wider than the empty part on the left, but the two halves are equal. In the center, a Helmholtz variant: the two sets of bars form superimposable square blocks, but the left-hand square looks elongated in the vertical direction, and the one on the right looks elongated in the horizontal direction. Below, an even more effective variant developed by De Savigny (1905). The two figures fit into identical squares.

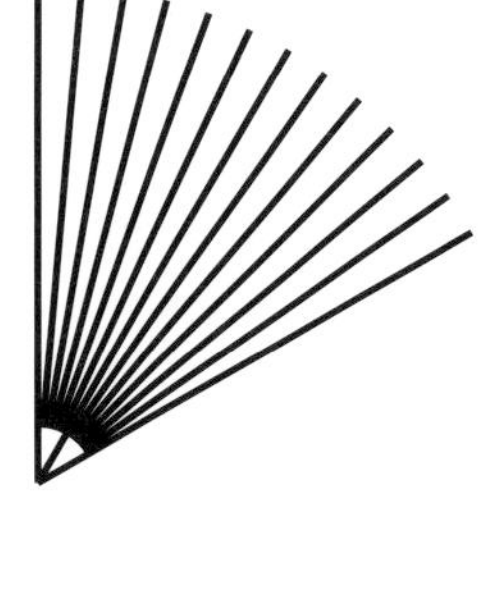

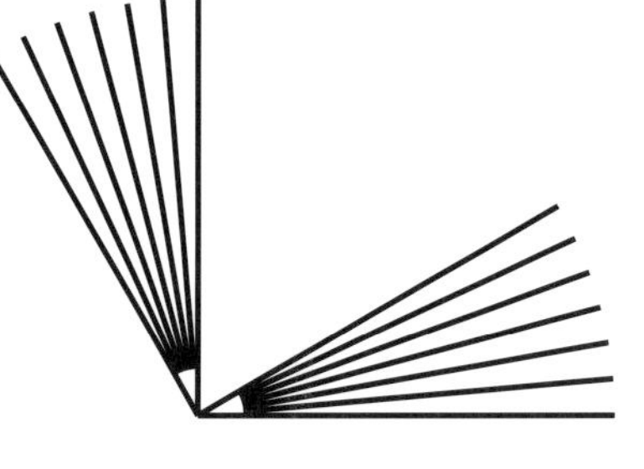

FIGURE 12-9. Tolansky's fan. The upper subdivided angle looks more closed than the empty lower angle, although they are equal. However, the stripes generally cause an effect of expansion perpendicular to their general direction, and we would thus have predicted the opposite effect.

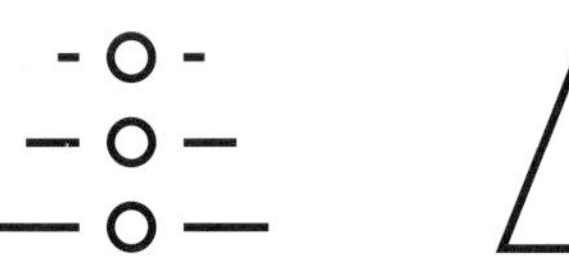

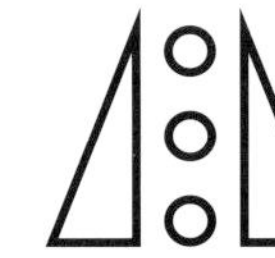

FIGURE 12-10. Titchener's illusion. At the top, the internal circle on the left surrounded by large circles looks smaller than the internal circle on the right surrounded by small circles. This illusion is generally presented as an effect of contrast. On the contrary, I see it as an effect of "normalization" in which we would tend to enlarge the small figures and diminish the large ones. In this interpretation the effect of diminishment on the left applies equally to the internal circle and to the large circles surrounding it. A similar effect is observed in the figures in the middle, with subjective contours. The lower figure on the left is like those in the center, but here each circle is flanked by only two elements, which can thus have their effects only according to the horizontal that carries them. The upper circle looks a bit larger than the one on the bottom. The same effect is observed in the lower figure on the right. The triangles appear to separate going from the base to the tip. These figures are at the intersection of Titchener's and Ponzo's illusions on the one hand and Bourdon's on the other (Figure 12-11).

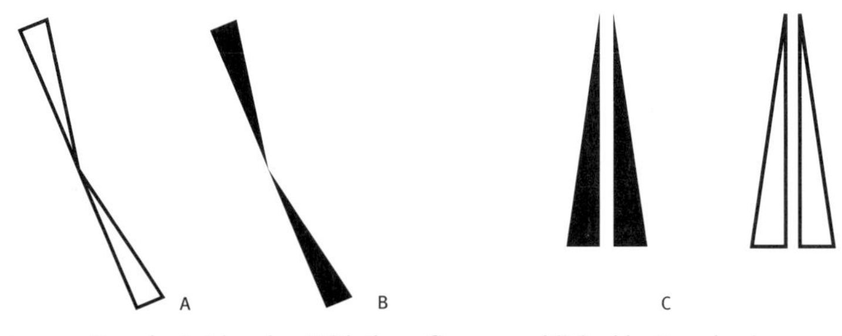

FIGURE 12-11. Bourdon's triangles. With these figures, published by Bourdon in 1902, France made its official entry into the field of geometric illusions. In (A) and (B), the left sides of the triangles are aligned, but they seem to form an angle toward the right. In figures (C) and (D), the space between the two triangles has a constant thickness. It nevertheless appears to shrink toward the points of the filled triangles and toward the base of the empty triangles. Before this illusion, others had been mentioned by Frenchmen as "scientific recreations" (see, for example, Figures 10-3 and 12-12) but received no official recognition.

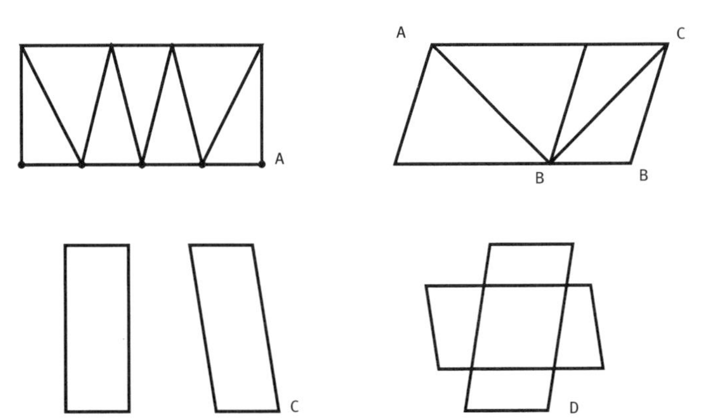

FIGURE 12-12. Parallelograms. Figure (A), the illusion of Dr. Fee, chief physician of the Eleventh Corps in Nantes, is perhaps the first geometric illusion discovered by a Frenchman (*La Nature* [1888, second half-year]: 287). The lower side of the rectangle is divided into four equal segments, but the segments in the center look smaller than the ones on the sides. In (B), Sander's parallelogram (1926), the diagonal AB looks larger than BC. In (C), Botti's illusion (1909), the parallelogram seems to have a larger surface area than the rectangle to its left, but their surfaces, calculated according to the formula base x height, can only be equal. In (D), two superimposed parallelograms are oriented like Shepard's tables in Figure 3-4. Is there still an illusion?

an analysis diametrically opposed to the usual one. The latter sees a contrast effect in these illusions: the large circles "oblige" the circle they surround to look smaller, and the small circles oblige the circle they surround to look larger. In my analysis, on the contrary, there is a tendency to equalize the two groups of figures; the one that takes up the most space is reduced and the other one is enlarged. The internal circle is thus reduced when surrounded by large circles, and enlarged when surrounded by small circles. This reasoning is symmetric with that given for the Müller-Lyer illusion.

For several illusions I have compared the errors I make depending on whether I look at them with my right eye or my left one. Within 10 percent, the errors measured are the same, for all the illusions studied, and

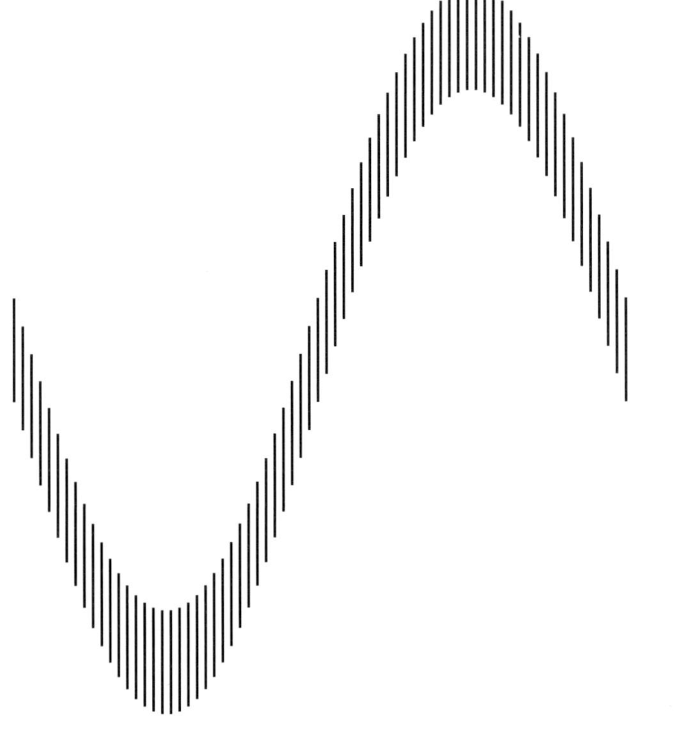

FIGURE 12-13. Day's sine illusion. All the vertical segments are equally long, but they look much shorter halfway between the tips.

for all figure orientations. But as these errors, in amplitude, are small and on the average only 2 or 3 percent of the proportions of the figures, the result is that the two eyes provide concordant information to the 2 to 3 percent level! We started with studies of the illusions thought to show the unreliability of the senses and we arrive at a result that shows them carrying out some extremely careful work, although its purpose is not yet clear to us.

One can find it obvious that the two eyes provide such mutually corroborating pieces of information for stereoscopic vision, in assessing relief, makes use of tiny differences between two images. It hardly matters that the two images are distorted, provided that they are distorted consistently. I will even go further: before the images can be compared, they must be

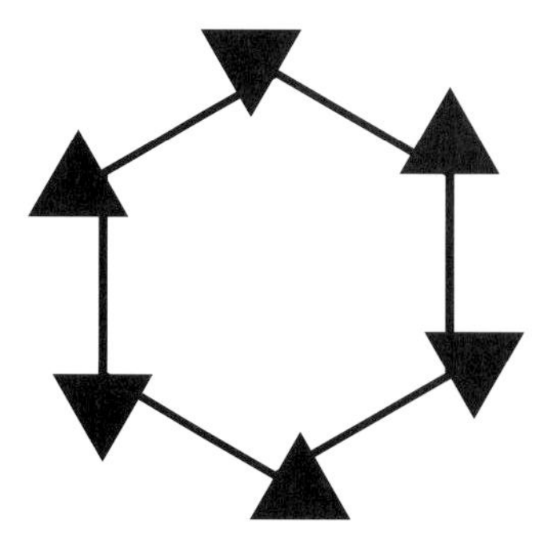

FIGURE 12-14. Mask effects. Two of the many illusions discovered by the Italian school. Right, Gerbino's illusion. If extended, the straight segments would form a perfect hexagon whose points would be masked by the triangles; but they look out of line. Below, one of Kanizsa's illusions. The diamond hidden by a rectangular mask looks, incorrectly, smaller than the other one.

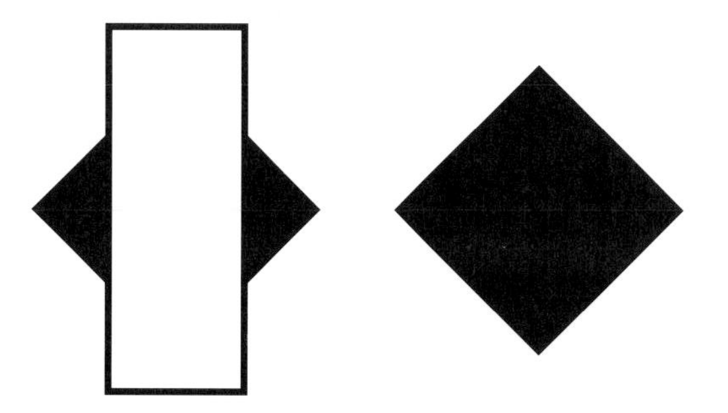

put in a common format and therefore corrections must be applied to them. It is as if, having convened a forum of experts who speak various languages, we get them to agree on a common terminology that was not strictly that of any one of the languages, but into which one could easily translate the terms of each.

Other problems of harmonization still arise. We can look at a scene with a fixed eye or with exploratory movements, by successive sampling. With a fixed eye, the image follows the law of linear perspective: a rectangle seen head-on projects as a rectangle, and the straight lines remain straight. With a mobile eye that turns toward various points of a scene, the image follows another geometry, that of curvilinear perspective. This is the usual way of looking, and yet the result resembles the one that would have been obtained with a fixed eye; the brain has thus distorted the image to obtain this result.

Finally, there is a way of grasping the image when everything is fixed, and a way of analyzing the scenes through motion: when one moves, the whole image changes, and the motion analyzers know how to extract the information by "direct" procedures that do not boil down to the comparison of successive images. Here too, when one is fixed and then moves, the geometry does not change; the two modes of visual apprehension of the scene are in perfect agreement in their conclusions.

13 ILLUSIONS AND CULTURE

We have become unable to grasp all that is strange and contrary to nature in a photograph. Anthropologists who have gone off to observe tribes that are unacquainted with writing and have had no contact with Western civilization tell how a member of a tribe reacts when shown a photo for the first time: he weighs it in his hand, turns it over, and does not know what to do with it. They explain to him that it is a portrait of his son: "Here's the head, there are the legs. . . ." A child who happens by understands what the photograph is, and he completes the description with: "Here is the nose, there are the eyes. . . ." The adult then realizes; he too sees it. This experience alone is enough to educate him; from then on, he interprets other photos without difficulty.

Our ability to understand photos is more surprising than it seems at first glance. In nature, light intensities that reach the eye contain enormous contrasts—between, for example, a white wall reflecting the harsh light of the sun and a tree branch deep in the shadow made by other branches. From a photo we receive the light reflected by the various parts of the paper's surface, light or dark, and the extreme values may stand in the ratio of 1 to 100—that is, a hundredth of the time it takes when viewing a scene in nature. Therefore, many shades are lost, and the photographer must give up reproducing all the details, either in the dark areas (which will then look deep in an impenetrable shadow), or in luminous areas where the harshness of the light washes out the details just as radically.

Although the range of brightness in the photos is low, they look sharply contrasting to us; half of a face black in the shadow connects with the other half white in the light, but they are separated by a too-strong dividing line. In a natural situation, according to Ernst Mach, shadows help us to grasp the shapes of objects; the shadow as such is not seen, but is automatically transformed into a depth cue. Conversely, in a photo, the shadow is like paint covering a part of the scene. Painters and engravers tone down the shadows and light so that their paintings or etchings are readable all over their surface.

Starting with the problems posed by perspective, printing and photography took divergent paths. Perspective tells us how we would see the world if our eyes were fixed, which is never the case. For example, a sphere to one side of a scene is seen in perspective as an ellipse. That is indeed what we see in photos, except that we mistakenly attribute the elliptical aspect to the poor quality of the lens, believing that it distorts the scene at its edges. Painters take it upon themselves to represent spheres, even at the edge of the picture, by circular discs, and we see no error in this.

The science of illusions is subject to two opposed temptations. On the one hand, neurophysiologists try to explain the workings of the mind on the basis of measures of electrical current in the neurons of the nonhuman brain. When a perceptual phenomenon is discovered in man, the researchers attempt to construct an experimental situation allowing the phenomenon to become detectable in the monkey, and they measure the electrical signals produced by its neurons to find out whether the mental phenomenon in humans would have a neuronal equivalent. If so, they conclude that the monkey and man perceive in the same way. A diametrically opposed "culturalist" tradition is prevalent in the human sciences. Anthropologists go into tribes that have had little contact with our own civilization, subject the natives to questionnaires meant to determine how they react to some of our own images, and conclude that, decidedly, people of different cultures do not see in the same way. But the subject matters of their cross-cultural studies do not match up with ours except when it comes to colors.

Neurophysiologists tell us that "color constancy" (the fact that an object's color is correctly identified under strongly colored lighting conditions, which are liable to be misleading) is observable in the monkey. Anthropologists teach us how much the classification of colors (or surface qualities) differs from one culture to another. Thus, certain peoples have only three terms for color—black, white, and red.

Neurophysiologists tell us that "subjective contours" (Chapter 8) are constructed very early on in the process of interpreting an image, before the interpretation of shapes, and they know what neurons are devoted to the process in the brain of monkeys. Proponents of a cultural approach—or here rather a cognitive approach—suggest that the surface perceived on the basis of subjective contours is a solution to a problem of interpreting the image, a solution linked to the meaningful forms in our culture—for example, the letters of the alphabet or, more generally, forms that require higher processing. This is the case of those complex three-dimensional shapes that arise from a small number of inducers on a pair of stereoscopic images (see Figure 8-7). Kanizsa's position in this debate is subtle. To say that we see a subjective triangle in Figure 8-2 because such a triangle accounts best for the particular arrangement of cues and the gaps between them merely shifts the problem. For him, the fundamental question is:

> What is a gap for the perceptual system (for the brain)? Or, in other words, why are certain figures complete or incomplete for the perceptual system? I have the impression that Gregory does not ever ask himself this question, in any case not explicitly, whereas for me it is a central problem.

Geometric illusions have brought about the same kind of split. For the present, neurophysiologists do not have a great deal to say about it. Specialists in the automatic treatment of images have taken over from them. In order to extract the forms present in an image, these specialists search for universal processing algorithms (filtering, contrast enhancement, and so forth). The application of these procedures to raw images produces revised images that are sometimes slightly distorted, hence the hope of getting from them a model of the genesis of the geometric illusions. However, these attempts stumble over an obstacle that cannot be circumvented: the Morinaga paradox. Geometric illusions sometimes take on conflicting appearances. In Figure 13-1, based on the geometry of the Müller-Lyer illusion, we see, consistent with the illusion, the distances between the points of outgoing angles

FIGURE 13-1. Morinaga's paradox. If we pay attention to the spaces between the angles, we see the outgoing angles as farther away from one another than the ingoing ones, consistent with the Müller-Lyer illusion (see Figure 12-5). But if, changing the criterion, we try to follow the positions of the tips of the angles vertically in any column, it seems that the ingoing angles take up more space than the others, which is inconsistent with the first illusion.

as larger than those between the points of ingoing angles. But if we turn our attention to the vertical alignment of the angles, we see them form a sinuous path where the deviations are contrary to those predicted by the size illusion. In short, the representation we form of Morinaga's configuration contains a few contradictions. In it, one distance may appear greater or smaller than another depending on the method of comparing them. This observation runs counter to a theory of illusion as an objectively distorted image, unless a different image treatment is proposed for each way of looking at an image. Attentional effects also play a role in processing images where the shapes of outgoing angles are superimposed on ingoing angles. One segment will then seem longer or shorter than another, depending on whether it is completed mentally with one type of angle or the other (Figure 13-2; see also Figure 13-3 and Figure 13-4).

At the other extreme, there are the cognitive-science explanations, such as Gregory's, of geometric illusions. According to him, the two members of the Müller-Lyer illusion make one think, in turn, of the projecting angle of a building façade and of the angles in a room viewed from within. One can get carried away with this game and find other correspondences between an illusion and a familiar configuration in our industrial-urban society (Figure 13-5). Thus the idea that tribal societies, living in villages without right angles, where the huts would be round, would be less sensitive than we are to geometric illusions. I think that the popularity of Gregory's theory with anthropologists stems primarily from the fact that it legitimated their travels, just as the space agencies like to put "biological experiments" aboard their space satellites which have no real interest for them other than to justify their space shots.

Following a study by Pitt Rivers in 1901, an investigation of great scope was conducted in the 1950s by anthropologists and psychologists working in exotically named tribes—the Banyankole, the Fang, the Songe, the Hanunoo, the Suku, and so forth—living in twelve African villages and a town in the Philippines. Their susceptibility to four illusions was measured: the Müller-Lyer illusion, Sander's parallelogram, and the L and T variants of the horizontal-vertical illusion. The same tests were given to three "Western" samples, two in the United States and a third in South Africa. This study was

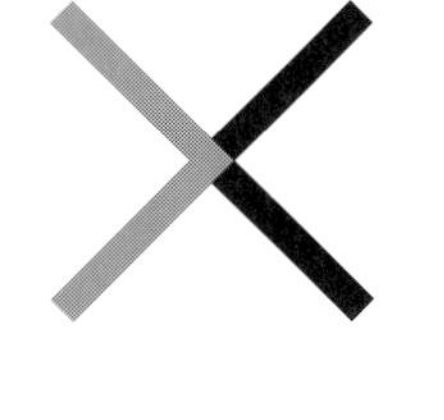
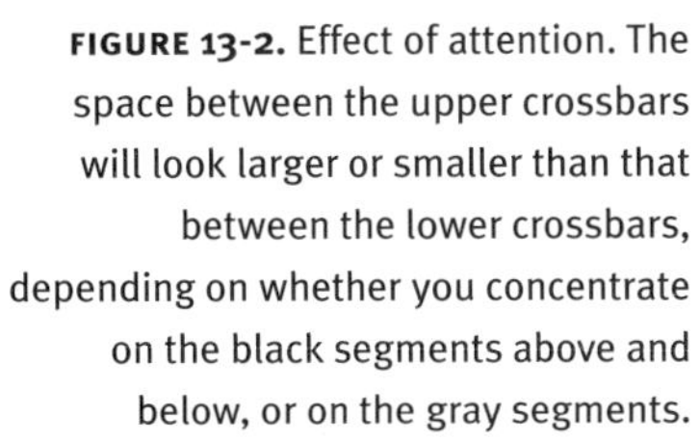
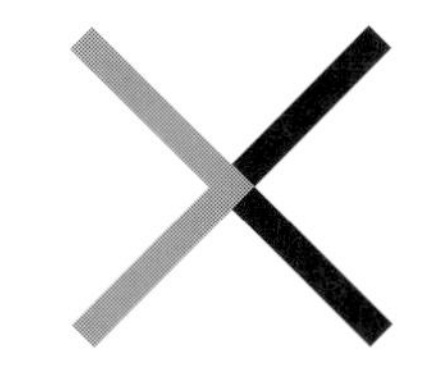

FIGURE 13-2. Effect of attention. The space between the upper crossbars will look larger or smaller than that between the lower crossbars, depending on whether you concentrate on the black segments above and below, or on the gray segments.

FIGURE 13-3. Shepard's uncertain circles. If you ignore the parallelogram and the ellipse indicated by the letter C, the figure is organized into two parallelepipeds that come toward the front; as in the figure on the left, the ellipses marked A can be seen as nearly circular, whereas the ellipses B will looked very elongated. If, on the contrary, you concentrate on the upper part of the figure to the point of seeing a single vertical block as in the right-hand figure, the ellipses B can look practically circular, and the ellipses A will be horizontally elongated (example adapted from R. N. Shepard, *Mind Sights*).

coordinated by Segall, Campbell, and Herskovits. They concluded that in relation to the Occidentals, the men living under tribal conditions were more susceptible to the horizontal-vertical illusion, less susceptible to Sander's parallelogram, and much less susceptible to the Müller-Lyer illusion.

One of the major difficulties of these studies has been that the responses obtained have depended on how the individual understood the question, which was conveyed to him or her by an interpreter who could adapt it. Despite all the precautions taken, it may be that the differences that showed up during these studies were unrelated to the character—urban or rural—of the environment.

Futhermore, it is probably worthwhile to determine whether, within our culture, important perceptual differences exist between one individual and another.

Do people see and hear in the same way? It is surprising to learn how belatedly the discovery of perceptual differences among individuals occured. Of course, it has long been known that some people had defective vision while others were deaf. But more qualitative differences, although noteworthy, have only recently come to light. For example, color blindness in its various forms—such as defective vision leading to confusions of red and green—was first discussed in a scientific paper in 1790—even though the condition affects 10 percent of the population in our societies, and must have afflicted at least 5 percent in antiquity. Jacques Rohault, in his *Treatise on Physics* of 1671, suspected its existence, for he himself had an *acquired* deficiency in color vision:

> I would, however, dare to affirm that, as it often happens that the same meat has very different taste for two different persons, so two men may

FIGURE 13-4. Kanizsa's fluctuating gray. This figure can be interpreted in two ways. We can see in it a transparent gray rectangle set on a white rectangle, the black circles being part of the white sheet. In this case, the gray looks relatively light. We can also see in it a white rectangle with circular openings revealing the black rectangle beneath, with the gray rectangle being inserted between the two sheets. The gray then looks darker than in the preceding case.

have very dissimilar feelings looking in the same way at the same object. And I am all the more persuaded of this as I have had an experience that is peculiar to me. It happened that I injured my right eye by looking through a telescope for more than twelve hours at a battle between two armies a league away. My vision is now in such condition that when I look at yellow objects with my right eye, they no longer look as they once did nor even as they do now with my left eye. And what is admirable is that I do not see the same variety in all kinds of colors but only in a few. For example, green seems close to blue when I look at it with my right eye. This experience of mine leads me to believe that there are perhaps men who are born with and have life long the condition that I now have in one of my eyes, and other men who see as I do with my other eye. Neither they nor anyone else can recognize it, because each person is accustomed to

FIGURE 13-5. Perspectivist interpretation. According to Emile Javal, "We will find it hard to believe that lines ab and cd are perfectly equal, for we know that the armoire is not as high as the room." The lines ab and cd completed by the corners in which they end form Müller-Lyer-like configurations. The configuration ab is encountered for vertical lines at a distance, whereas the configuration cd is more likely to be seen on closer objects. Hence a perspectival explanation of the illusion: we are in the habit of judging as longer lines like ab and shorter lines like cd, "and we keep this habit when the reason for its existence no longer holds"—for example, in the Müller-Lyer illusion (*La Nature* [18 January 1896]: 111–12).

naming the sensation produced in him by a certain object with the name already in general use but which, however common for the various sensations that each person may have, does not make its meaning any less ambiguous.

Equally surprising is the way in which humanity could miss out on an entire dimension of vision, stereoscopic vision. The fact that with two eyes one can better assess depth than with one eye had indeed been noted, here and there, but without the reason for it being very clear.

In audition, the observation of qualitative differences was more belated. In the first decade of the nineteenth century, Wollaston discovered that in a group of persons of equal sensitivity to low-pitched sounds, some of them heard the highest-pitched sounds, and others did not hear them at all; that thus certain persons with normal hearing do not hear the chirring of crickets or the very high-pitched squeaks of bats. Today we know that temporal resolution is also variable:

> Leipp tells us such a subject will clearly distinguish impulses separated by a few milliseconds; another, on the contrary, will merge them into a uniform jumble! Anyone who has exceptionally fine temporal resolution will be interested for example in the very rapid songs of certain birds (the lark, the robin, and so forth) whose "melody" appeals to him or her, whereas another person will find this song uninteresting. Similarly, a certain musician will be attracted by particular musical instruments whose very brief transients he perceives (harpsichord . . .), while another does not even hear them because his ear is not "quick" enough.

Often cultural differences are superimposed on these individual differences. In acquiring language, the child learns to group certain sounds and to treat them as variants of the same vowel or the same consonant; these groupings vary according to the language. A well-known example is that of the Japanese, who do not differentiate between "l" and "r." In the West, rhythmic patterns last a few seconds, and we cannot recognize a longer rhythmic unit, but certain traditional kinds of music employ rhythms stretching over twenty seconds or more. We are impregnated with music that follows a tonal system with constraining rules, and that has a circular organization (like the hour marks on a watch). The result is that certain pairs of sounds, separated by half an octave, can be, depending on the notes and the person, perceived as ascending or descending (chapter 7). According to Diana Deutsch, such personal biases would be connected with the acoustic categories formed during language acquisition.

One of the rare occasions on which people were led to notice their differences and to speak of them happened when, in 1992–1994, a new kind of image, the autostereogram, appeared on the market and generated keen popular interest. Was the relief in a given image perceived as strong or weak, embossed or hollow? Were the surfaces seen as continuous or on successive planes? Was the image easier or harder to see than another (see chapter 4)?—all this varied from one individual to another. This was probably not a big surprise, for we always find large individual differences in laboratory experiments, both qualitative and quantitative. When the classic geometric illusions are measured seriously, we find that even individuals with the same cultural background react very differently to various aspects: strength and sometimes even sign of the illusion, variation with the figure's orientation, relative susceptibility to the different types of illusions.

For many experiments in experimental psychology, if we wish to have a clear idea of the situation, we must test "many" subjects; by "many" I mean around twenty. In a sample like this, it is rare for sizable individual differences not to appear. But these are taboo. More precisely, the scientific journals readily publish studies finding differences between men and women, or between populations with different skin colors, but they do not accept the mere observation of variability within an undifferentiated population.

Some researchers gloss over these differences by publishing articles in which they use only two or three experimental subjects, generally the authors themselves. The result is findings with illusory consistency. People jest at times about primitive societies that have only three words for numbers—"one," "two," and "many." One very media-oriented journal peddling predigested scientific data has published articles on several occasions about perception in which it spoke of "subjects" in the plural, without indicating the number (which I assume equaled two), no doubt fearing that the reader might otherwise form an accurate idea of how frivolous these articles were.

Another classic magic trick to cover over qualitative differences is to substitute statistics for them. Instead of saying there are 10 percent left-handed people and 90 percent right-handed in a population, a published article will report measuring an "index of laterality" of 0.9 with a variance of 30 percent.

When a subject takes the same tests several times in a row, his responses change. The illusion does not always diminish; it may grow stronger. At the start, the subject has not yet found "his" best way of apprehending the figure. The answer to the question asked—does A look larger than B?—is not really clear. After several trials, a way of seeing the figure and of gauging A in relation to B takes shape. Then his responses become quick and repeatable, without necessarily involving a diminishment of the illusion.

We make fun of the drawings produced by other cultures and about the "errors of perspective" made by painters or draftsmen of foreign civilizations. Even if it is not explicitly stated, this filters through in treatises on the history of (Western) art, which accord inordinate attention to the discovery of the laws of linear perspective and to its use in painting. At the same time, we consider the distortions that Cézanne introduced in representing the body or the manipulations of images by contemporary graphic artists to be the daring of genius. We find it hard to admit that systems of representation foreign to our culture also have their daring. In the mandalas made by Tibetan monks, the characters are depicted on the four sides of the picture, their heads toward the nearest edge, and the picture can be viewed from all four sides. This conception of the image is completely defensible, independent of religious symbolism. Moreover, in European paintings on ceilings, we sometimes find this distribution of characters in all orientations, with their heads toward the center.

Reversed perspective is observed in certain cases. Remember Euclid's remark: "If a sphere has a diameter less than the distance between the eyes, the viewer will see more of it than a hemisphere." This observation has more impact when we look head-on at a small object with plane faces—for

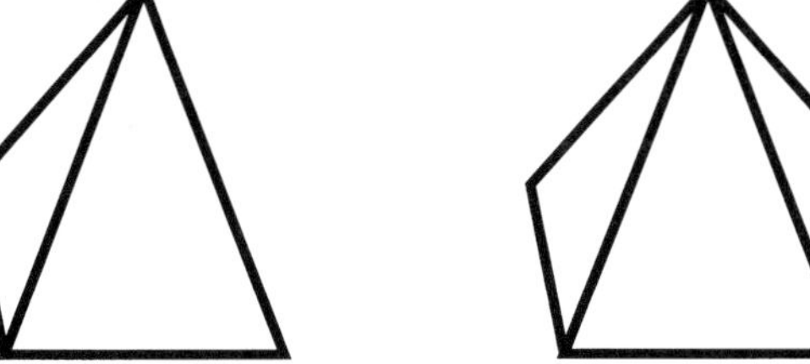
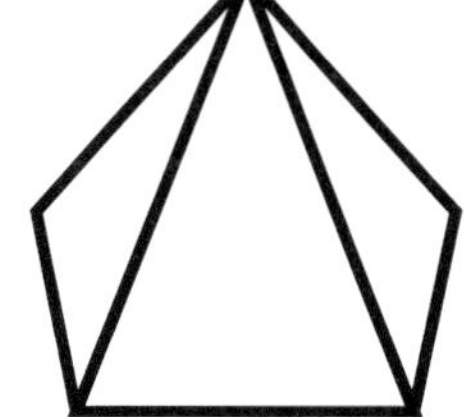
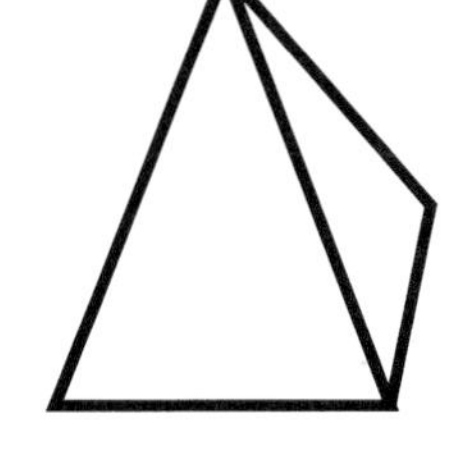

FIGURE 13-6. Inverted perspective. A small pyramid (with a base less than the distance between the eyes) would be seen by the left eye as shown in the left-hand figure, by the right eye as shown in the right-hand figure, and by both eyes together as shown in the figure in the middle, where it is seen with three faces at the same time. Such a representation is contrary to the laws of classical perspective.

example, a cube or, better, a pyramid. The left eye sees the front and the left side, the right eye sees the front and the right side, and the two views connect to give an image in reversed perspective (Figure 13-6), in which one sees the two sides and the front at the same time—which will never appear on a photo. The frequent use of reversed perspective in the pictorial art of the Far East perhaps follows from a very elementary technical constraint: painting on fans requires objects to be gathered closer together toward the bottom of the fan, thus to represent the front as narrower than the rear. Once you are accustomed to this mode of representation, perhaps you develop a taste for it and find advantages in it. Perspective shows only what is possible to see from a single viewpoint. With reversed perspectives, the front of a house can be shown at the same time as what is happening on its left side and on its right side. In one and the same fresco—I am thinking of that by Ramayana in Bangkok—you can see painted scenes where the perspective is reversed as well as others, still representing complicated architectures, where the perspective treatment is close to European norms: a sign that in the former scenes the inversions were deliberate.

14 ILLUSIONISM

The most surprising thing in certain magic tricks is that the means used to take in the observer are so simple and are often known to him. For example, in the trick in which you see a speaking human head set on a three-legged table, the head's owner is squatting, unbeknownst to the audience, with his head going through an opening in the tabletop. But in front of him are mirrors that hide him and create the illusion of a floor continuing under the table.

Yet the power of mirrors to deceive is well-known: often, in public rooms, a large mirror covering one wall "enlarges" the room and gives the illusion that it extends beyond the wall. Seated, one can be fooled, and catch on to the trick only upon leaving. The same effect can thus be used in a sensible and legitimate way as an aid to decoration or in a mischievous way by a magician to show things that appear to be impossible. Finally, these effects have also been used perversely by the oracles of antiquity and by today's magi and sorcerers to produce belief in their supernatural gifts or their connivance with the gods.

As regards illusionism, certain tricks require exceptional dexterity and preparation of the magician, others use rigged equipment, and I regret not knowing how to judge the true difficulty of each trick. Certain demonstrations of "mental telepathy" between a host and a medium demand memory and great presence of mind. The two partners have agreed on a code: the exact wording of the question asked by the host and possibly his intonation provide cues to the medium that steer him toward the right response. What do I have? What am I seeing? What do I have in my hand? Tell me. . . . What do you see? And the like. Each of the different wordings sends a message. We are taken in by the speed of the answers in the face of the very short questions, and we have trouble imagining that they could convey secret information. Other demonstrations of mental telepathy, alas, use stage sets with mirrors or stooges hidden under a trap opening who whisper the answer (Figure 14-1).

With miniature radio sets, "mental telepathy" can be performed by partners in closed and separate rooms. The impresario of one wizard of mystification, Uri Geller, perfected a radio set lodged in a live tooth, with which a medium could receive information in the form of little jolts transmitting a message in Morse code. You can also buy rigged dice containing a small transmitter with which the fake sorcerer can guess the number that has come up, although the die is out of his sight, in another room. Geller succeeded in guessing the numbers thrown on his own dice, but refused to do the experiment with dice provided by others. One trait of those who earn their living by demonstrating "paranormal gifts" is that they reserve the right to announce that they are not having a good day, that their occult powers are failing them, and so forth. Everything that these "mediums" perform is also carried out by professional magicians who never give poor excuses and who present their tricks every evening without choosing their audience.

Many tricks are based on substitutions of objects. The magician shows a rope, cuts it into short pieces, then with a gesture makes it look completely reassembled. The viewer tries to imagine that the rope was cunningly faked, that it was made up of ends that fit into each other, whereas the magician has simply cut the rope, which was not the rope shown at the end. Although aware than the magician is able to bring about such substitutions, we reject this explanation. Very early in childhood, we understood that a face going peek-a-boo always belongs to the same person, that the plaything that reappears is the same as the one that had been hidden. The hypothesis of the permanence, stability, and continuity of objects is perhaps one of the strongest in perception and one of the main resources of magic tricks.

One surprising piece of fakery is performed onstage: the reassembling of an actor cut into pieces. Following an explosion, a person is cut into eight pieces. Another person gathers up the arms, thighs, torso, and head in a basket. Next, the magician takes a leg, goes to the back of the stage, sets it down, proceeds likewise with the other parts of the body, and reconstructs the injured actor. At

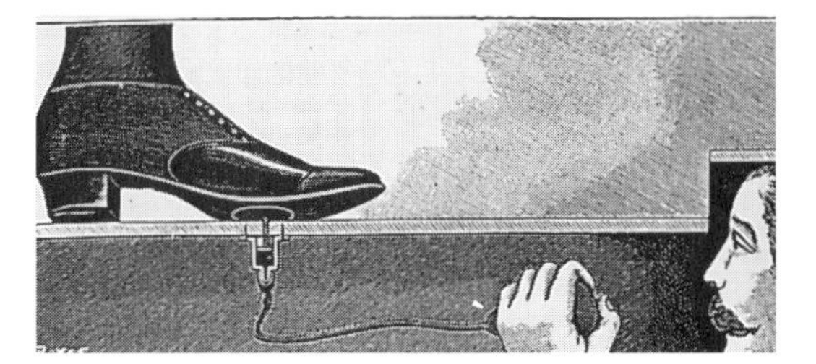

FIGURE 14-1. Mental telepathy in the theater. This performance was based on a trick: "A small pneumatically operated piston in a rubber tube was installed in the sole of the accomplice's shoe. A stooge, placed under the stage, keeps his eyes on the blackboard on which a member of the audience wrote numbers, and squeezing the bulb fixed to one end of the rubber tube, he displaces the piston beneath the shoe. The accomplice is thus informed, by a code previously agreed upon, what she must say" (*La Nature*, 1899, first half-year, p. 111).

the end of the operation, the reconstructed body sways back and forth as if it could not stand by itself, but then begins to walk as before. This trick depends on substitutions. The method used in the nineteenth century is shown in Figure 14-2. Today performers must have more powerful methods, for I have seen puppets transformed onstage into actors at stunning speeds.

In this trick or one of its variants, there is a time during which the flesh-and-blood person is enveloped by his inanimate replica. A supplementary refinement can be used to play with the face. A person in the flesh moves about on the stage. The viewer does not know that his face is in fact covered with a mask. When the person goes into a box and sticks his head out through one of the openings, it is actually a second mask, identical to the first one, that comes out. It is then easy for the magician to make an empty space appear beneath the head.

In movies, many tricks use substitutions. Often, the nonhuman hero (King Kong or E. T.) comes in two sizes: a miniature form from twelve to twenty inches (thirty to fifty centimeters) high used for most of the scenes, when he is seen whole, and a larger form for certain parts of the body. When King Kong carries the heroine in the palm of his hand, the film uses a hand that was built on a large scale. In adventure films, the hero has stand-ins for the perilous scenes, a stuntman or an inanimate dummy. In scenes where someone is beheaded, the rolling head is rarely that of the actual actor. I also remember the movie *La Dernière Femme,* in which a French actor cut off his penis with a power saw; a year later, in *Rêve de singe,* one found this organ attached at the place intended by nature. There was certainly faking in either the earlier film or the later one.

Often, in illusionism, "the illusion is in the proof." I thus recall a show where a magician rose up in the air and soared with the ease of a bird. The idea that came to mind was that he was suspended, activated like a marionette by invisible wires, which in our time would be nothing extraordinary. The audience, however, saw the magician whirling his arms around, holding hoops. The viewer could not help thinking then that the magician could not be suspended by wires, for wires would have blocked the movements of the hoops. The magician's real work, in my opinion, was in his handling of the hoops.

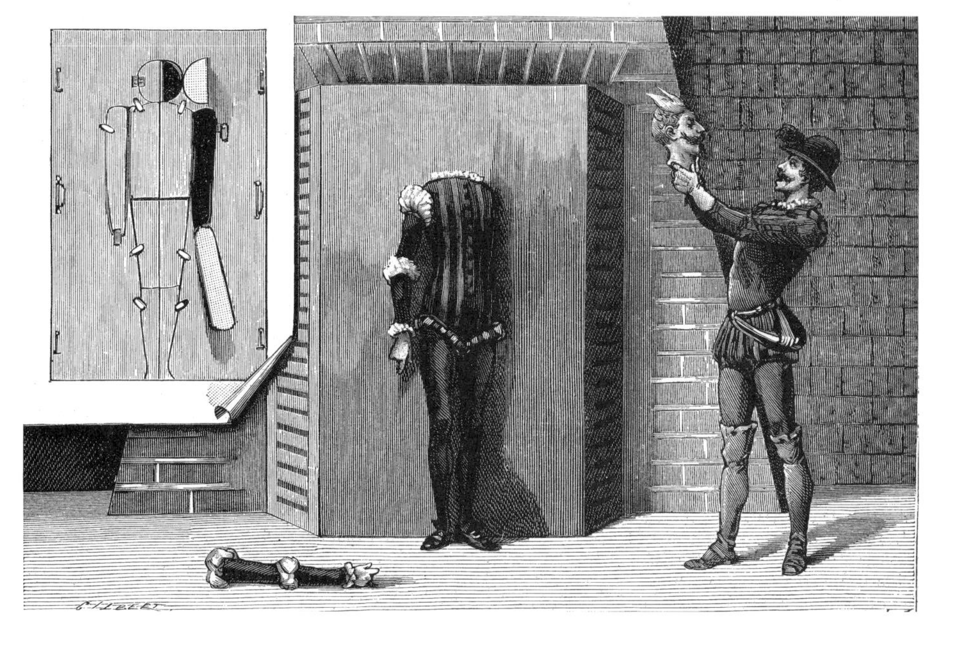

FIGURE 14-2. The actor sliced into pieces. Parts of a dummy are taken to the back of the stage and set against a black curtain. Behind the curtain is an actor who, through openings in the curtain, inserts his legs, then his arms, into the corresponding parts of the dummy. "Thus, piece by piece, the true actor replaces the dummy; he is not exactly at ease during the operation, because he is obliged to hold back the upper parts of his body until the end . . . When the head is finally placed, and the substitution is complete, he must play at being the dummy for a few instants and give the impression of something inert with unstable balance" (*La Nature*, 1890, second half-year, p. 96).

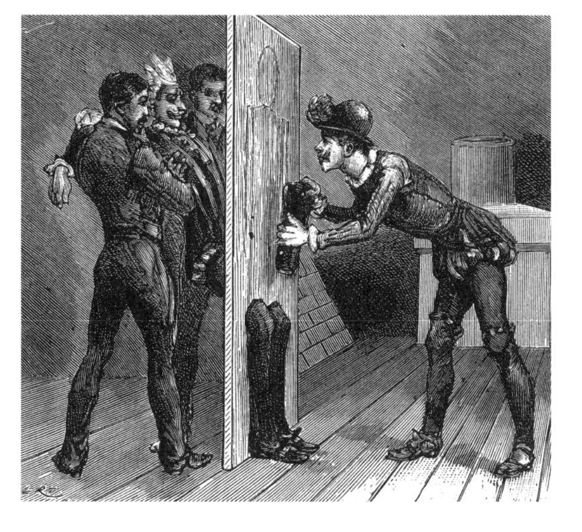

Many tricks are done without any apparatus for faking and depend on the magician's manual skill; even an informed eye does not see his gestures, as expressed in the saying "The hand is quicker than the eye." I recently had a chance to analyze the recording, reproduced on videotape, of a trick performed by a professional magician. We saw him take out of his pocket a piece of floppy plastic a few inches long—a child's blue balloon—then blow into it. When the balloon reached its full size, about the size of a head, he tied it at its end. Next, holding the balloon by the tied-up end in his left hand, he grasped a dagger in his right hand and made a quick move to burst the balloon. The balloon burst and we then saw that it contained a live blue bird beating its wings. On reflection, we think that the bird could not have been in the balloon in the first place, that there must have been a high-speed substitution. How this substitution had been accomplished was revealed by replaying the videotape frame by frame.

On television, the images on the screen change fifty times a second, a given image being displayed in two consecutive frames, one for the even lines, the other for the odd lines. A complete image thus extends over some forty milliseconds. Here is how the critical part of the magic trick went (Figure 14-3):

First image (time T). The magician holds his dagger in his right hand, ready to strike at the balloon. The balloon is held in the left hand laid flat against the chest.

Second image (time T + 20 milliseconds). The balloon is burst, the knife is no longer visible, and the right hand that held it reaches toward the lapel of the jacket from which it pulls out a blue bird.

Third image (time T + 40 milliseconds). The bird emerges, held in the right hand, which throws it toward the left hand (the one that held the balloon): the right arm is along the body, where we would expect it to be at the end of the knife blow.

Fourth image (time T + 60 milliseconds). The bird flies to the left hand.

Everything was performed in the twenty to forty milliseconds taken by the second image. In the third image, the blue patch of the pigeon is already in the space previously occupied by the blue patch of the balloon.

Having dissected the trick, I played back the recording several times at

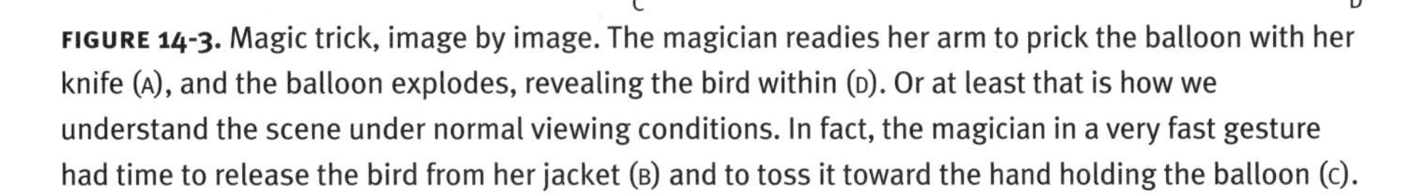

FIGURE 14-3. Magic trick, image by image. The magician readies her arm to prick the balloon with her knife (A), and the balloon explodes, revealing the bird within (D). Or at least that is how we understand the scene under normal viewing conditions. In fact, the magician in a very fast gesture had time to release the bird from her jacket (B) and to toss it toward the hand holding the balloon (C).

normal speed: the right hand making the knife stab seems to make a quick, continuous gesture; at this speed, I never managed to spot the gesture of extracting the bird. Incidentally, we know that if one inserts in a film an image completely unrelated to the action, the viewer is not aware of this image. It has often been claimed that the information contained in the odd image was nevertheless taken in unconsciously, and that it then influenced the viewer's behavior, which would make possible a form of hidden advertising or political propaganda. The effectiveness of these subliminal images has been much exaggerated: it appears that a precise message like "drink Coca-Cola" sparks a vague desire to drink, but this desire does not settle on any particular product. It is clear that an unexpected image that lasts only a twentieth of a second, inserted in the stream of images, leaves no conscious trace; in this interval the gifted magician can perform a crucial gesture, under the public's very nose.

Many techniques are used to give the illusion of objectivity as much in economic or political information as in the reporting of scientific work. Statistics are particularly effective to this end, and charlatans do not fail to make use of them to give a scientific polish to their frauds. The most common technique of statistical charlatanism consists of providing "significance levels." It is commonly said that a particular result is significant at one in a thousand or one in a million. By that, we understand that there is one chance in a million that the observation is due to chance, and since this is very unlikely, it must be due to what the authors think it is due to.

For example if I toss a coin in the air under ideal conditions, theoretically it has one chance in two of landing tails up, and one chance in two of landing face up. Suppose I toss it a great many times and determine that the chances of finding "tails" occur in at least 51 percent of the cases. This probability is one in twenty for 10,000 tosses, one in two and a half billion for 100,000 tosses, and falls vertiginously when the number of tosses increases further. We see how this applies when transposed to political life. Having had to reach a decision by referendum on an interbank agreement they had not read, the citizens of a banana republic approved the measure by 51 percent, which earned them congratulations for their intelligence from the head of state. Statistically speaking, the result was highly significant—of the sort that, according to the calculation of probabilities, could not even happen once in fifteen billion years throughout the universe, had it been due purely to chance alone. Nevertheless, any person with common sense knew that the result would have been different had the weather been rainy instead of sunny on voting day or if the bank of a neighboring country had raised interest rates by a quarter of a point instead of lowering them by a quarter of a point, and so forth. In fact, all that statistics prove is that there is a disparity relative to a perfect theoretical ideal. But they in no way prove that the disparity is due to the factor that was supposedly being tested.

Often scientists let themselves be deceived by statistics. In journals of experimental psychology we increasingly read articles in which the authors say that they compared one situation with another and found a significant effect at a level of so much out of a million, without taking the trouble to specify what direction the effect was in. It is as if on the day after the referendum, the French found headlines such as: "A loud and clear result"; "The French did not vote at random"; "The turnout in areas is significantly correlated with voters' preferences," and so forth, without saying whether the "yes" or the "no" had carried the day.

Another technique of statistical illusionism consists of making "significant correlations" appear in large tables of data. For example, I examine a sample of two hundred individuals and I notice a number of their characteristics: the incidence of this or that illness, sexual behavior, skin color, social success, and so forth. For each individual, I also determine the characteristic of a great many genes. Let's say that for each gene I have two possibilities, labeled + and –. Having done a systematic study of a great many genes, I can construct a double-entry table of the following kind:

	TYPE OF GENE PRESENT				
	GENE *1*	GENE *2*	GENE *3*	GENE *4*	GENE *5*
Skill at pinochle					
Individual 1 (brilliant)	+	–	–	+	+
Individual 2 (brilliant)	–	+	+	+	–
Individual 3 (no-hoper)	–	–	+	–	+
Individual 4 (no-hoper)	+	+	–	–	+

From a table like this I can infer that there is a correlation between skill in pinochle and gene 4, since the skilled individuals have the + variant and the weak ones the – variant. What must be understood is that even if the pluses and minuses are distributed completely by chance, if you have a large enough number of columns you will indeed find one where the distribution of pluses and minus is, by chance alone, correlated with one of the traits observed in the individuals. Thus it is that the journals of predigested science regularly announce, with statistics at the ready, the discovery of the gene for this or the gene for that. The publishers of these journals know that these announcements will not stand up for long, that the correlations will crumble with the next installment of results, and that they will then invoke a "second gene" for this or that, then a "new factor" influencing the activity of the genes, and so forth.

So why is this cheap science always pushed to the front? The role of these journals should be understood in relation to the daily press and other media, of which they are the regular suppliers. In the choice of articles published, the commercial criterion of resale to the media comes well before the scientific criteria.

Political, scientific, or religious debates are often distorted according to an immutable principle: one brings together the person who is wrong, who is a hardened demagogue, and whose cause one secretly espouses, to face an opponent who is right but who does not know the case well enough to counter his adversary on precise technical points. Take the case of the charlatan who claims to transmit thoughts at a distance. A newspaper that claims to be objective, well-balanced, reader-respectful, and nonpartisan will put two discourses in opposition: that of the charlatan who claims to have abilities not explained by physics, and that of his critics: academicians or Nobel Prize winners who will bring out their authority, express their righteous indignation, say that they cannot give any credence to a phenomenon so manifestly opposed to the most sacred laws of physics, and the like. The reader to whom the two contradictory discourses have been served up will not fail to congratulate the newspaper for its remarkable objectivity. The only one who will not be given the floor is the professional magician who "knows the trick" and could perform it without further ado for the public. Had he been allowed to speak, the reader would understand everything right away, and there would be nothing left to write in the next few days on this subject. The whole art thus consists of getting the charlatans to speak on the one hand and the distinguished scientists to speak on the other, provided the latter have nothing relevant to say on the subject. But it sometimes happens, alas, that an independent journal comes along and lets the cat out of the bag.

15 ARTIFICE AND CONVENTION

Look in a noisy bar at a group of customers packed together in front of a television set tuned to a football game. On the screen, seen from the side, human forms just over an inch (three centimeters) wide and less than six inches (fifteen centimeters) high fight over the little brown spot that is supposed to be the football. Caught up in the action, the customer imagines himself on the edge of the playing field, between the players on the field and the clamoring public in the stands. The familiar voices of two sports announcers heighten the illusion of being present in the stadium. Why are such poor images that are displayed on a small screen able to rouse such powerful emotions?

The old images, icons or painted portraits, fascinated the public for whom they had been made. Then, because of technical progress or simply excessive familiarity, these images lost their power of illusion, of substituting for reality. Seeing them, we find it hard to put ourselves in the shoes of the viewers of old, refuse to believe they were able to have extraordinary sensations of spatial depth from easel paintings, or, later, from ordinary photographs seen through a magnifying glass (Figure 15-1).

Accustomed to seeing images on certain supports or in certain frameworks or in a particular style, we can be trapped by innovations: at the start, the image using a new procedure (for example, in painting, chiaroscuro or perspective) acts on us through surprise, because it incorporates aspects of reality that we were not used to seeing in an image. Later, it produces a perverse effect. The new images become the norm and are regarded as faithful, legitimate reflections of reality, including what is conventional in them, stamped by the technique of the moment. Finally, once the procedure has become part of our lives, the image loses a great deal of its power of illusion. Thus, in the early days of movies, the representation of a train coming into a station could throw the audience into a panic, which could not happen today, even with the progress in the technology of filming.

The presence of a frame delimiting the image makes its artificial character manifest. When we become oblivious to the frame, by looking at the image through a tube or through a hand cupped like a tube in front of an eye, we have a better experience of depth; we can do this experiment at the movies or with still images (Figure 15-2). For the latter, the grain of the paper may give a clue to the flatness; the best effect is obtained with slides, seen with a viewfinder against the light of a light box for viewing negatives.

"Perspective cabinets"—some date back to 1750—were boxes in which one could see through a peephole a complex scene, including furnished rooms opening onto other rooms, which provided a surprising sensation of reality. The inner faces of the box were painted, and the box contained some real objects and mirrors to create subsidiary volumes, which were actu-

FIGURE 15-1. Depth with a single eye. This apparatus has only a magnifying glass and a mirror at forty-five degrees. It was reputed to provide "a very remarkable appearance of relief." Engraving from *La Nature*, 1886, first half-year, p. 176.

FIGURE 15-2. Claparede effect. Observe this image through a narrow tube—for example, a rolled-up sheet of paper or a cupped hand held against the eye. The design will take on relief similar to stereoscopic relief.

ally seen in reflection. Surfaces seen as vertical could well have been painted beginning on the floor and extended onto a wall. Here too, the fact that the eye, riveted to the peephole, is immersed in an image whose edges it does not perceive strongly contributes to the effect of relief (Plate 3).

Around 1900 there was a proliferation of apparatuses giving the illusion of depth, but which were not based on the stereoscope—for example, the "monostereoscope" and particularly the *vérant* and the "synopter" made by Karl Zeiss. The monostereoscope or the zograscope (see Figure 15-1) was primarily a large magnifying glass through which one looked comfortably with both eyes at a flat photograph reflected by a mirror. The *vérant* was a box with an eyepiece for looking at slides. The image was sent to infinity, and the eye kept a single focus while exploring the image. The focal distance of the eyepiece was adapted to that of the image, assumed to be taken with a standard camera; thus the real perspective was reestablished. Finally, the outlet pupil of the eyepiece coincided with the center of rotation of the eye, avoiding the "keyhole" effect. The synopter, patented in 1907, was a binocular apparatus with mirrors (Figure 15-3), made for observing a single image. This image was received by the two eyes as if it was seen by a single eye located between the two. The photographs seen with the synopter were perceived with a relief that, for some observers, was even more vivid than true stereoscopic relief. According to Koenderink, on the other hand, the natural scenes seen with the synopter took on a strange flat appearance that was unlike what one would see by closing one eye. Moreover, under these conditions, size constancy would be practically abolished.

At the end of the eighteenth century the Scottish painter Robert Baker began creating giant pictures, "panoramas," paintings done life-size on large cylindrical surfaces that the viewer was to admire by taking a position at the center (Figure 15-4 and Figure 15-5). The dioramas developed by Bouton and Daguerre starting in 1823 used primarily tricks of lighting to heighten the illusion. The viewer was still at the center of a cylindrical room, on a turning platform, and admired successively the scenes presented in separate cabinets. The painted canvas was at the back of a dark cabinet, the light came through a space between it and the ceiling, and the viewer saw the whole thing at a distance, through a small opening. In 1826, Vergnaud wrote:

> The amazing illusion of the panorama has nothing comparable to it unless it is that of the diorama; a partially animated painting in which mechanical means and a particular mode of lighting create a sky that is in

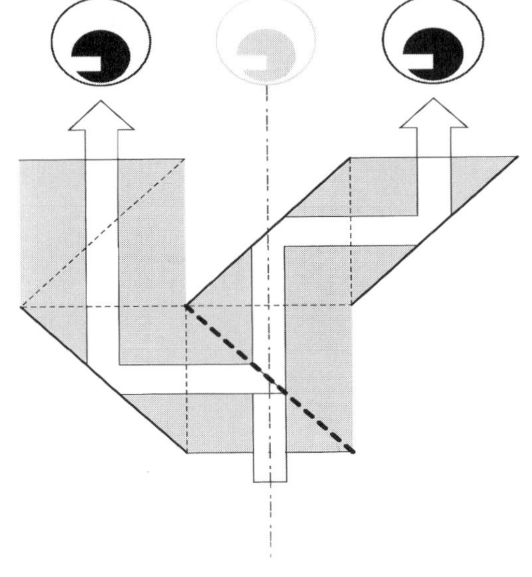

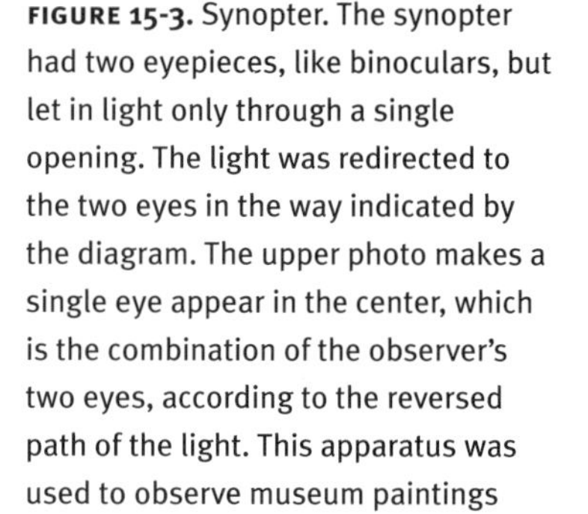

FIGURE 15-3. Synopter. The synopter had two eyepieces, like binoculars, but let in light only through a single opening. The light was redirected to the two eyes in the way indicated by the diagram. The upper photo makes a single eye appear in the center, which is the combination of the observer's two eyes, according to the reversed path of the light. This apparatus was used to observe museum paintings and provided strong illusions of relief.

some manner moving, and changes of light and shadow that in nature are brought about by the imposing and regular march of the stars and the wandering course of the clouds.

A few years later Daguerre invented the "double effect" diorama. Two pictures were painted on the two sides of a transparent support. Lit from in front, the first painting was visible; lit from behind, the other. (A similar trick is used in the theater: a piece of scenery painted on a semitransparent support stops the gaze when lit from the front. When, suddenly, it is lit from

FIGURE 15-4. The Coliseum. This panorama, built in Regent's Park in 1829, was a gigantic painted canvas representing London as seen from the dome of St. Paul's Cathedral. The viewer, brought up by elevator to the galleries at the top of a cylindrical tower (pictured on the left), had a plunging view of the paintings that covered the wall of a cylindrical building. Engraving from Rudolph Ackerman, *Graphic Illustrations of the Coliseum* (1829). See also Kemp (1990).

FIGURE 15-5. The Avenger. This panorama set up on the Champs-Elysées in 1892 represented a naval battle between the French and the British fleets off the island of Ouessant in 1794. The viewer, standing on a platform, witnessed a shipwreck of the ship *The Avenger*, painted on the cylindrical canvas. He was supposed to be on the bridge of the brig *Le Courrier*. All around, waves were formed by sturdy canvases stretched over intertwining boards. The platform was driven by a pitching movement 20 inches in amplitude. "The effect obtained is quite enough to give the illusion of the movement of the ships represented on the panoramic canvas. The waves around the brig also seem to heave, and the viewer thinks he sees the glorious debris of *The Avenger* swallowed up right before his eyes." From *La Nature*, 1892, second half-year, p. 129.

the rear, the decor disappears, and the scene extends instantaneously toward the rear, where one sees new walls and new furniture.)

Panoramas and dioramas also used various processes of trompe l'oeil used in perspective boxes, stage scenery, and replicas from wax museums. One of the techniques was to have elements with one portion real and solid, and the other portion, an extension of the first, painted on canvas. The solid half was itself repainted in the style of the half on canvas. Another process was to place elements of decor on successive vertical planes, recessed in such a way as to create effects of parallax.

The cinema also attempted various gigantic effects, including complete panoramic shots, in which a scene was filmed by eight cameras covering the 360 degrees of the horizontal plane. The projection was done on the wall of a cylindrical room with the viewer at the center. This technology has less effect on us than the panoramas and dioramas had on the audience of a hundred fifty years ago. The current process of projection on a giant hemispheric cupola or the related one of projecting on a giant flat screen, vertically elongated, also produces excellent effects. In both cases, the upward extension of the image is important for the illusion, whereas the tendency in films had been to favor, for the great-spectacle films, the panoramic screen extended horizontally. Television chose to develop a format with a width/height ratio of 16/9, much less square than the current format (4/3). No study proves the superiority of the 16/9 format. I estimate on the contrary that to bring about a strong sense of reality, it is preferable to extend the visual field upward.

Besides using projection on a giant screen, moviemakers wanted to add a supplementary element of realism, stereoscopic relief—a particularly costly and largely illusory refinement. I had an experience with this technology in an amusement park devoted to images. Each viewer was furnished with a pair of liquid-crystal glasses. The images for the left eye were projected alternately with those for the right one. The left and right lenses of the glasses grew darker in synchrony with the images in such a way that each eye received only the images intended for it. The sensation of depth was excellent, but it didn't have much to do with this feast of technology. It was due primarily to the fact that the glasses, which were opaque on the sides like diving goggles, shrank the visual field on the edges, and thus caused the effect of immersion in the image favorable to the sensation of depth. The viewer would have had the same sensation looking at any film through a tube that was blackened on the inside.

Often, these films for the giant screen are documentaries that introduce a region of the world that the viewer has not had an opportunity to visit. The camera is mounted in a small airplane or helicopter, and the audience discovers the landscape as if sitting next to the pilot. Sometimes the aircraft flies hedgehopping over a plateau, comes to the edge of a precipice and dives, and the audience feels in their stomachs the sensation they would have had in real life.

Movie technology ratcheted up one notch in "kinesthetic cinema." The viewer is still led to explore as if he were aboard an airplane, rocket ship, or roller coaster. The scenery is sometimes natural, sometimes created with synthetic images. To add to the realism of the bodily sensations, the seat he is sitting in is subjected to little movements synchronized with the events shown on the screen. The process is effective, and the public screams when the images and the seat give sensations of sudden acceleration or braking before an imminent collision.

During the showing of one such movie, we were led out of a tunnel that opened onto a yawning chasm, but instead of falling, our vehicle went off to explore other areas. How could the bodily sensation of a fall have been transmitted? A person in a free fall is subjected to the force of gravity as at every other moment of his life, and does not have the sensation of being sucked downward. On the contrary, he has the feeling of floating, which dreams sometimes convey well. I told myself that day that the kinesthetic cinema had met its limits.

Let us review what could, in an ordinary projection of a movie, warrant the term "illusion." First, a physiological illusion: the sensation of continuous projection. The large reels on which the film is wound go around at a regular rate, but the portion of film that goes in front of the light projector makes stops. The image shown on the screen comes from a frame of immobilized film. Between the projection of this image and the next one, the flow of light is cut off, and the room is plunged into darkness. The consequent flickering, barely detectable, has been eliminated nowadays: modern projection apparatus sends light between the frames. The succession of snapshots on the film is interpreted by the viewer as one uninterrupted scene. During a shoot with low-sensitivity film, each of the images of a moving vehicle is sometimes blurred. This does not prevent us from seeing it as sharp—a fine performance of the brain, which manages to create clear views out of fuzziness.

Lighting conditions are constantly manipulated in movies. We enter a dark jail cell, and in a few seconds the darkness vanishes. In a dark cave, the hero takes out his lighter, and the little flame seems to light up many square feet of the walls. In interior scenes, the lighting is almost always uniform and the characters do not make a shadow. Or the actor heads toward the window from which light comes, in principle, and we clearly distinguish his shadow on the wall where the window itself is located, proof that he had strong lighting from the rear. Often, outdoor night scenes are filmed in broad daylight using the process known as "day for night." The contrast is pushed to the maximum between the actor in the foreground, wearing light-colored clothes and flooded with artificial light, and the set in the background. A red filter is affixed to the camera in order to darken the sky in relation to the rest of the landscape and to soften the details, and the film is developed so that only the foreground is light.

When the camera explores a scene, we are in thought in the position of the camera in that we follow its forays in every direction; however, the image is straight ahead on the screen.

The tracking-in shots, in the old style, done without changing the camera's focal length, corresponded well to natural vision. With those shots, changes in the image produced parallax cues that made it possible to interpret the scene in depth. Today, there is a tendency to replace the tracking-in shot with a gradual lengthening of the focal length (the "zoom shot") without moving the camera. The result is that the central part of the image grows until it occupies the whole field, without any change in the relations among the parts of the image. The parallax cues are lost, and one sometimes experiences a compression of the scene. Naturally, when a long focal length is used, the scene is compressed for a reason of pure geometry explained in Chapter 3. Vehicles filmed with a telephoto lens on a highway take on a ghostly look.

When the camera moves sideways, at a fixed focal length, it also changes the relations between parts of the image, it masks or unmasks backgrounds, and also generates parallax cues. But if the camera pans across the space in a pivoting movement (or "panoramic shot") the relationships among the scenic elements are preserved. Here, it is the panoramic shot that is closest to natural vision, except when one sees a landscape through the window of a moving vehicle. Finally, we have acquired great tolerance for the medium-range framing that shows people cut off at the waist by the lower edge of the image, and an equal tolerance for the close-up framing that shows heads cut off at the neck kissing other heads.

Despite its unnatural changes of focal length or its unrealistic movements, the camera gets forgotten. For example, the hero and heroine are flying in a little prewar crate, seated one behind the other without much space, as in a kayak. The next shot shows us, without transition, the mountains toward which the airplane is rushing, and we cast an eye on them with the same anxious look that's on the actor's face. Although everything was filmed from another plane, the illusion is twofold: looking at the plane from the outside, we proceed to the inside; and furthermore, immediately take the pilot's place.

On the other hand, television reports do not fool me. When they take us to uninhabited lands, and when they show us one man alone in the face of nature, making strange gestures in a perilous situation, and when the narrator glorifies this solitary effort, I cannot help but think of the film crew gathered around the man, of the acrobatics performed by the cameraman with one hand clinging to his equipment, the other hanging on for dear life, and his legs searching for footing.

The management of time generates many oddities. Everything appears to be following in chronological order whereas the crew went to one place, filmed all the scenes set there, then moved elsewhere for another group of scenes. Within a scene, a minor exchange of three sentences between hero and heroine could be shot in ten takes, and the actress perhaps refreshed her lipstick three times before each attempt at kissing.

During editing, depending on the order in which various sequences are presented, different stories can be suggested. Kuleshov and Pudovkin had filmed an actor in close-up who had been asked to be expressionless. The shot was then inserted in a sequence just after some images of a bowl of soup or those of a dead woman or of a child playing. Seeing these sequences, the viewers credited the actor with expressions corresponding to the context.

Despite all opposition, the miracle of the cinema works, and all implausibilities are accepted. Taras Bulba, Genghis Khan, and Cleopatra express themselves in English. In the films of the 1950s and earlier, Indians, Arabs, and even blacks had brown skin, but their faces, if we disregard the makeup, were typically those of American whites. In the epics, when the arms of the extras were raised to approve of the Roman emperor, it was not rare to see some who wore wristwatches. In adventure films, it is still customary for the explorer to go off on an expedition for several months without extra underpants—but the woman explorer never forgets her depilatory cream.

16 ILLUSIONS OF THE MEMORY & THE MIND

It is extremely rare that two human beings, taken at random on the planet, agree about politics, philosophy, sex, or religion. In the life of a couple, each of the spouses goes through most of his or her life laying the blame on the other about how to drive or to behave, about decisions on purchases, or the proper place for things in the home. According to this observation, half of humanity would be in error, not only about the questions of doctrine and of conceptions of life, but also about all the little details of practical life.

One illusion, possibly the strongest of all, is the one that makes us believe that we have a direct hold on reality. The work of interpretation conducted by perception never comes to light and leaves no other trace than its final result. Although the natural world that surrounds us is revealed to us by way of electrical signals exchanged among neurons, we have the impression of contact with it at a distance. The tennis player feels the ball at the end of his racket; the handyman feels the resistance of the screw at the tool's tip. In the laboratory von Békésy created tactile localization "at a distance" by transposing to the domain of touch the principle of stereo sound or vision. Subjects to whom one transmits two synchronized vibrations to the ends of the index finger and the middle finger respectively are able, after practice, to localize the sensation at a point intermediate between the two fingers, even if the fingers are spread apart. In an experiment of the same kind, a subject who receives synchronized vibrations on his two thighs, just above the knees, manages, after practice, to localize a vibratory sensation between his two knees, although they are spread apart.

With a less demanding technique Ramachandran recently created an illusion of contact at a distance, known as "the Pinocchio illusion." A subject is seated, blindfolded. Ramachandran takes him by the hand, and has him tap the nose of a student seated in front of him. At the same time, he himself taps the subject's nose. After thirty minutes the subject feels his nose growing longer.

This experiment comes within the framework of a more general initiative to understand the illusions of amputees. We have long known that people who have lost an arm or a leg strongly feel the presence of the missing limb, sometimes for years. Some of them have sensations of a moving "phantom limb"—shaking hands, for example—and other people have the sensation of a limb paralyzed in an uncomfortable position. One amputee, who had the illusion that his phantom arm stretched out at a right angle to one side, instinctively turned sideways each time he went through a door. Ramachandran thought that these acquired paralyses perhaps originated in a lack of visual stimulation: the brain would believe the arm to be paralyzed because no visual signal could confirm the proper carrying out of its directives when it ordered the arm to move. He thus used a ruse to create the missing visual sensation. He placed the subject with an amputated left arm sideways to a large mirror; that is, with the nose against the edge of the mirror. In this way the subject saw, on one side, his normal right arm and, on the other, the image of his right arm in the mirror, producing the illusion of a left arm. Ramachandran then asked the subject to make symmetrical movements with the two arms, like a conductor leading an orchestra. In a flash, the amputee felt his phantom arm unfreeze. This method succeeds, it appears, in 50 percent of the cases.

It remains to be understood why the brain maintains the fiction of the phantom limb. Why does it feel sensations of heat or cold, of pinprick, of contraction, of pain? The old idea was that the nerve endings present in the tip of the stump continue to send messages, which are interpreted as if they came from the end of the phantom limb. According to Ramachandran, it is a phenomenon of synesthesia, similar to the one of colored vowels. One area of the brain responsible for the management of signals coming from the amputated limb is annexed by a center that manages another part of the body: the left cheek would thus take over the left-arm area. Small real stimulation of pinprick, moisture, or heat on the left cheek would then be translated into corresponding illusions in the phantom limb.

The sense of exteriority extends to our purely internal productions such as dreams and hallucinations. It takes a paradoxical form with afterimages. When a person looks at an object briefly illuminated, after the end of the flash of light, he forms a negative afterimage of it (Chapter 10). If he then closes both eyes, the afterimage remains visible and goes to the inside. In an investigation conducted by Harvey Carr with his students, he obtained two types of descriptions: some students reported confidently that they saw the afterimage in their eyes, generally on the internal side of their eyelids; most of the students were reluctant to say that they saw the image in front of their eyes, but admitted that it *appeared* to be outside their eyes, while insisting that the image had to be in reality either on the retina or in the brain.

Whatever our philosophical or religious position on the relation between mind and body, we live this relation as obvious: there are situations in which we think of ourselves as a mind that makes decisions and gives orders to the body, and other situations in which we feel ourselves in the grip of bodily demands of hunger, thirst, sleep, or itchiness.

We have the sense of acting rationally according to what we have seen or heard. Very often, however, the decisions are made unbeknownst to us. Orders for movement are given well before we become conscious of a sensation. The well-known extreme case is that of a person who, having recklessly put his hand near a fire or a very hot surface, takes it away in an instinctive movement and feels the burn afterwards. In everyday life, we perform many gestures without our awareness: tapping a foot, scratching, holding one's chin, nibbling at something, passing one's tongue over one's teeth, and so forth. And when we begin to hum a tune that keeps running through our mind for no apparent reason, from where did the order come? And if we tell ourselves to stop, why is this order not carried out? When the gesture is intentional, and it is performed well in conformity with the goal assigned, we do not realize all of the orders that were computed, organized, and sent to the muscles, to put the decision into practice.

If we think that the mind controls the body, we must admit that it is content to give general directives, and that most decisions are taken on its authority, by a "second mind," a part of the brain that acts first and then informs its hierarchical superior. This is evident when you make an error while typing some text on a keyboard: you know you have made a mistake before seeing the result displayed in print.

I recently found myself in a situation in which, it seems, two or three areas of the brain were fighting over the control of my fingers. I was performing a very simple experiment with surprising results: the "weighing of numbers." It was a matter of saying which of two numbers between 0 and 9 is higher than another. The two figures, taken at random, were shown on a computer screen, and we measured the time taken to indicate the larger number. Intuitively, one expects that "4 greater than 3" has the same obviousness as "7 greater than 2" and that, consequently, the response time will be the same for all the comparisons. What Moyer and Landauer showed in 1967, however, is that the closer the figures are, the longer is taken to respond, and the more mistakes are made. I repeated the experiments and noted, for one thing, that my errors were conscious (I knew that I was making a mistake at the time I was pressing on the key) and, for another, that the mistakes often happened in consecutive pairs. It was as if somewhere in my brain a center was mistaken in the comparison, and had given to my finger the order to strike the wrong key, whereas another center of the brain, better at arithmetic, instructed my consciousness but did not know how to control my finger. As soon as I made a mistake, my watchfulness increased, and I was convinced that the next time I would answer correctly, but then there I was with my finger declaring 2 is greater than 9. It was as if the center controlling the finger movements had said to itself: "Since you are not satisfied, sir, I shall do the opposite."

Conversely, the desire of men and women for one another is often presented as bestial relations that come from the body, where perceived nakedness and the contact of skin trigger the most primary instinctive reactions. All of that involves highly intellectualized processes, however. I make an exception for the sense of smell: we are said to have a second sense of smell, the "glomerulo-nasal system," entirely dedicated to sexual odors, which does not manifest itself to us by any conscious smell, but acts surreptitiously on the bodily expressions of desire, and makes us attribute great beauty, or extraordinary mathematical insight, to a person whose odor, in truth, our second olfactory sense appreciates.

The parts of the body considered erotic have varied with eras and cultures. Our current European culture has focused, for marketing purposes, on a part of the female body known to keep its qualities of appearance and suppleness for a very long time. Its principle of construction "in the form of an imperfect sphere, split in the vertical direction, offering rather exactly the aspect of a two-leaf clover," according to the description by G. Courteline in *L'Article 330*, poses a few physical and perceptual problems that warrant our attention. Consequently, its power of attraction could be more intellectual than physical. Besides its twofold convex and concave character, the variations in its form according to one's point of view and especially the occlusive aspect creates surprise. An occlusion almost always indicates the interposition of one object in front of another. The signs of occlusion trigger the formation of a subjective contour (Chapter 8) that leads the observer to prolong the hidden surface in imagination. The psychology of perception has also not failed to emphasize that most of the objects of daily life like tables or chairs are kept in equilibrium thanks to four points of contact with the ground, or, exceptionally, three points. Other familiar spherical objects (marbles, balloons) keep in balance with a single point of contact. But the doubly spherical objects in the form of a dumbbell that can rest in equilibrium on the ground with two points are rather a rarity.

There is a kind of error of memory that we can produce with certainty. In an experiment done in the United States, a list of twelve words was read to a group of subjects: for example, "slumber, tired, rest, night, comfort, noise, eat, bed, snore, dream, awake." The subjects were distracted for five minutes, and then were asked to recall all the words they heard. On the average, the subjects—psychology students—remembered seven or eight words, but 12 percent of their words were intruders that were not on the original list. The intrusions were not just any words: they were words related to the words on the list, words that the list words were able to evoke, consciously or not. In one of the experiments done with forty-one students, there were forty intru-

INTELLIGENCE MEASURED. After ten years of thorough fine-tuning, an objective, reliable, and universal test of intelligence has just been finalized by a team of researchers of the well-known Massachusetts Institute of Technology, directed by Professor Royburger. This test, based on the comprehension of the rules of the game of baseball, has made it possible to establish that the Peuls . . .

INTELLIGENCE MEASURED. After ten years of thorough fine-tuning, an objective, reliable, and universal test of intelligence has just been finalized by a team of researchers of the well-known Massachusetts Institute of Technology, directed by Professor Royburger. This test, based on the comprehension of the rules of the game of baseball, has made it possible to establish that the Peuls . . .

FIGURE 16-1. According to Bourdon, on the double-spaced lines the letters look larger. The Peuls are a population in Africa.

sion errors, among them the word "sleep" twenty-seven times. In the set of responses, "sleep" was found as often as some of the words that were actually heard. The experiment confirms what one thinks of the organization of the memory: every stimulation received from the outside awakens everything in memory related to it. To hear or read a word is to substitute for a graphic or sound signal the word in memory that best corresponds to it. But at the same time that the right word emerges in consciousness, other related words are taken from the stock and are ready to surface as associations of ideas. Afterward it is difficult to distinguish a word really heard from a word evoked. Nevertheless, there are some qualitative differences between the two kinds of memories that are revealed when the subjects are asked a few questions:

FIGURE 16-2. (Opposite page) According to Day, a passage in the middle of a page looks larger than the same passage at the bottom of the page. This could be a variant of the second hat illusion (see Figure 3-3b).

Love finally explained. Already known were the genes determing sex and the hormones of desire. Only lacking was a piece of the puzzle for undertanding love in its totality. This has been done with the discovery of the neurons of love in the brain's antero-parietal area. "Even spiders need love," commented Professor Bethozard while presenting his latest results to the press.

Love finally explained. Already known were the genes determing sex and the hormones of desire. Only lacking was a piece of the puzzle for undertanding love in its totality. This has been done with the discovery of the neurons of love in the brain's antero-parietal area. "Even spiders need love," commented Professor Bethozard while presenting his latest results to the press.

(1) Were they sure that the word was on the list? (2) Did they *know* that the word was on the list or did they *remember* having heard it? (3) Did they remember anything in particular concerning the presentation of the word? On average, the subjects were more confident of having heard the words really on the list, they remembered them (in contrast to knowing they were there), and they had richer memories of the circumstances when they heard them.

Often the information that we receive is ambiguous, but we tend to settle on the first interpretation that comes to mind, and hold it as long as this interpretation does not become clearly untenable. We have all experienced a situation in which we had thought something and then, when the evidence that could disabuse us piled up, it was ignored or explained away in terms of the initial error. Afterwards, we recall well all those clues that would have put us on the right track. The cross-purpose dialogs in the theater are based on this principle.

Feature columns in the newspapers are sustained by absurd dramas based on the complementary principle of "I never would have imagined that" or "I would have thought of everything except that": a man is awakened by a noise in his home during the night; he grabs his gun and kills someone he lives with, for he never would have imagined that he or she could be strolling alone through the house that way in the middle of the night. In 1789, during the French Revolution, there was a famine, and Paris was starving. The astronomer Jean-Sylvain Bailly, who was mayor of Paris, related:

> 21 August. Provisions were so short that the life of the inhabitants of the metropolis depended on the mathematical precision, as it were, of our schemes. Having learned of the arrival at Poissy of a boat with eighteen hundred sacks of flour, I immediately sent from Paris a hundred wagons to fetch them. And lo and behold, in the evening an officer without authority and without assignment told me that, having come across these wagons on the road to Poissy, he had them turn back given that *he did not think* that any provisioned boat was waiting on the Seine.

Arago, who reported this passage in his biography of Bailly, added:

> Today, even after more than a half-century, we do not think without a shudder of this obscure individual who, for *not having thought* that a

loaded boat could be stationed at Poissy on 21 August 1789, was to plunge the capital into bloody chaos.

In logical exercises, people have a tendency to get stuck on what seems to be the obvious beginning of a solution, and they seek to complete it. They experience the greatest trouble in getting off the wrong track.

The inability to conceive of alternatives is also manifest in social relations, in the way in which a person is immediately classified by his facial features, by his way of dressing, or by a single sentence. In a second, at thirty meters' distance, a Frenchwoman, according to the traveler Landroy, "sizes a man up, before he has made the slightest gesture or spoken a word. She knows with an infallible instinct whether he merits attention or if it is proper pointedly to ignore him." That was a century ago, and I cannot say whether since then things have changed for the better or the worse.

In nature, an animal must decide in a fraction of a second. Is a form that it spots that of a predator or a possible prey? Its life and its survival depend on speed of judgment. Humans must also make up their minds quickly in their social lives. Shall I trust this stranger or not? The decision is tricky, for we live in a society of masks in which each person conceals his intentions and his faults. Thus we find it necessary to construct criteria of judgment that are worth what they are worth but allow us to avoid being frozen on the spot.

We generalize without rhyme or reason, based on our experience alone, by imagining that others operate as we do, from which arise all sorts of "illusions of penetration" concerning their thoughts, intentions, and motivations, foretelling what is liable or not liable to happen. In the long run, if we have given up trying to understand the inner workings of other people's minds, we have still acquired an ability to predict without too many errors how they will behave in this or that circumstance. And even then, we will occasionally have disillusions such as: "I would never have thought that of her [or him]," "I thought you were different from the others," and so forth. It seems to me that many of the faults that logicians find in our way of resolving their problems perhaps stem from the fact that our intelligence is primarily constructed for the treatment of social relations. In this domain, it is better to make a decision quickly, at the risk that it is not the best one, rather than to lose time assessing all the possibilities.

One very widespread illusion among French drivers is that they "gain time" when they bear down on the gas pedal to go faster. In city driving, because of the accidents caused by hurried drivers, the authorities have installed numerous red lights, which chop up traffic, and slow it down below the speed at which it would flow in a city without stoplights where drivers were well-behaved.

Suppose, moreover, that in going faster, the total duration of a trip is shortened by a certain time (t). This time t costs the driver a time T that is much greater. Indeed, the car going faster wears out sooner and must be replaced sooner. At best, the driver has gained a few hours but pays for them with hours of work. Even supposing that the balance sheet is positive, what use will he make of the so-called time gained?

We all lead a double life, the one awake, which leaves many impressions, and the one of dreams, in which we forget just about everything, although our experiences there are sometimes so unusual, and so personal. A savings of time, if there could be one, would be to enter into full possession of this whole hidden part of our life, to recover the best portions of it. In this sense, the best means of saving time would be to devote some of it to recalling precisely the best dreams.

The memory of the great moments of our life is another of our riches, lost if they remain buried. Chance encounters or a conversation sometimes awaken the memory of a whole part of our existence that otherwise might never emerge. We were thus richer than we thought. Penfield showed in the 1950s that electrical stimulation inside the brain, in the fronto-temporal cortex, reawakens fragments of life, with hallucinatory clarity. But he never established whether the scene rising to consciousness had been really experienced previously or not. In any case, it is conceivable that someday methods will be found that allow an individual to explore his memory systematically and to replay at will the best episodes of his life. Such a discovery would, however, endanger worldwide commerce, for it would make pointless the possession of most material goods: what use would it be to buy a luxury car if you could rent one for a day, store up the sensations, and play them back when you felt like it?

On the other hand, this discovery would provide untold advantages to the performing arts. Consider the ordeal of the performers, who spend hours every day practicing the pieces of music they will play in concert, up to the point of knowing them "on their fingertips"—that is, to the point of becoming able to play them while thinking of something else. The huge amount of time spent practicing will one day seem to us as pointless as the time an astronomer, in the era of computers, might devote to manually calculating the return of a comet. Someday, the musical interpreter, having found the interpretation appropriate for a piece of music, will know how to record the neural code concerning the corresponding muscular commands. At a concert, he will plug himself in to an apparatus that will supply the brain with impulses activating the desired finger movements. Freed of the physical performance, the mind will devote itself more completely to the bodily expression of emotion, which is so important in relating to the audience. The audience will enjoy symmetrical advantages, and through stimulation of an appropriate brain area, will enjoy the musical interpretation to the degree chosen or recommended, from casual pleasure to ecstasy.

Regarding illusions that apply to the present writer, I could claim the researcher's illusion (believing one has had a brilliant idea that is merely a recollection of something read long ago); the illusion of the historian of ideas (believing that ideas originated with the authors one has come across by chance); and that of the psychologist (generalizing to others on the basis of one's experience alone). The study of perception and that of illusions are so linked that it was difficult to treat the latter without doing a complete book on the former. Having expressed myself on perception rather completely in *L'Empreinte des sens*, I wanted to avoid repeating myself, so I have favored here themes that were not thoroughly explored in the first book.

The most complete book I know of on visual illusions is that of Robinson (1972). One may also consult Bourdon (1902), Coren and Girgus (1978), Kanizsa (1997), Lanners (1990), Lukiesch (1922), Metzger (1975), Da Pos and Zambianchi (1996), Tolansky (1964), and Wade (1982). For audition, I based my work essentially on the books of Bregman (1990), Leipp (1977), Warren (1982), and articles by Deutsch (1975, 1992), Risset (1986), and Wessel and Risset (1979) as well as the book of von Békésy (1967), which also treats tactile and gustatory illusions.

As a general rule, when an author is mentioned in the body of a book for a work or an idea, a reference to his or her name is given in the bibliography. The following notes add specifications and other information, chapter by chapter.

I wish to thank all those who have helped me in the preparation of this book either by providing illustrations—Isia Leviant for his original "Enigma" illusions (Plate 1), revolving cones (Chapter 11), and "pointed squares" (Plate 8); Gérard Bouhot for his photodocument (Chapter 4); Jan Koenderink and Andrea Van Doorn for documents on the synopter (Chapter 13); Martine Dombrosky for the drawings of Figure 14-3—and by clarifying various matters of bibliography—Nicholas Wade, always available to answer all questions; and Isabelle Leponce, archivist at the Center for the History of Sciences and Technologies of the University of Liège, who provided me with some old articles by Delboeuf and Plateau.

1. INVENTORY

This first survey in the domain of illusion does not aim to be exhaustive. I selected a few effects that are not laboratory constructs and are observed naturally provided that one goes to the trouble. The quotation about the green ray comes from a reader's letter to the journal *La Nature,* 1885, second half-year, p. 366. The quotation concerning Professor Lépine was published in *Nouvelles scientifiques,* 14 August 1897, p. 11, supplement to *La Nature.*

2. A BRIEF HISTORY OF ILLUSIONS

To go back to the sources, one can consult translations of the writings of Euclid (Burton, 1945), Ptolemy (Lejeune, 1956), El-Haytham (Sabra, 1989), and that of Kepler by Catherine Chevalley (1980). Ptolemy, about whom I have said nothing, discussed seeing simple or double objects depending on whether one uses one eye or two, and how the two eyes converge when focusing on an object, in front of it, or behind it. Moreover, he

TRAITE
DE
PHYSIQVE
Par IAQVE ROHAVLT.
TOME PREMIER.
Quatriéme Edition, reveuë & corrigée.

A LYON,
Chez Claude Galbit, en Belle-Cour, à la Maison de la Cage.

M. DC. LXXXI.
Avec Approbation & Permission.

spelled out the law of size constancy, and he described the fusion of colors with a spinning top (see Howard and Wade, 1996). El-Haytham described many perceptual effects rediscovered in the nineteenth century (see Howard, 1996). Among the books on the history of optics and of vision, let us cite E. G. Boring (1942) for the history of ideas, Polyak (1941) for the eye, and Lindberg (1976) for the formation of images. One will also find a detailed account of the old positions and extended extracts of old authors in Trouessart (1854), and anthologies of perceptual or illusory phenomena in Plateau (1878) and Wade (1996). Plateau's series of six articles is a gold mine for the historian of science.

Mariotte explained why normally we are not aware of the blind spot:

> My third assumption is that the eyes are extremely mobile, & what makes us see the exact detail of a whole object is the swiftness with which our eyes run over all its parts by direct vision, as when we read, for even though we perceive all the lines on a page at the same time using peripheral vision, we can only read them by looking successively right at all the words & nearly all the letters in each word, from which it follows that the habit of movement of our eyes hinders us from comfortably fixating at a determined point for a while.

He was thus not thinking that there was a "filling-in" of the blind spot (as in the experiment on the upper part of Figure 2–1), but rather that it was a recollection of what would have been seen previously in the same place.

Jacques Rohault is quoted very little, although he merits becoming a star in the French history of the cognitive sciences. One of the major problems confronting the search for the history of ideas of his time is that writers positioned themselves relative to Aristotle, but rarely cited their scientific sources, and Rohault displays this shortcoming. In any case, it seems to me after reading Malebranche (1712) and Rohault (1681) that the former, when he presented ideas about vision that were thought to be his own, very largely drew his inspiration from the latter.

The quotations from Lucretius and Seneca were taken respectively from the Garnier-Flammarion edition (1964) and the Belles Lettres edition (1929). As regards the blind spot, see for example the popularized article by Ramachandran (1992). For Rubens, see Topper (1984).

3. ONE ILLUSION HIDES ANOTHER

Few illusions can be analyzed in the pure state. The most striking ones combine several effects, so that to understand them well you become concerned, step by step, with all the known illusory effects. For illusions on terrain, one can consult Bourdon (1902), Carr (1935), and the article by Proffitt et al. (1995). To the effects cited of anisotropy (square-diamond, Hermann's grid) may be added the "oblique effect," according to which parallel lines are seen distinctly farther away when they are horizontal or vertical than when they are oblique. These orientation effects are curious: it is tempting to attribute them to an anatomical origin; the neurons concerned would be arranged in a square network, according to the two axes corresponding to the horizontal and the vertical. But neurons, like most cells, form rather hexagonal arrangements. Nevertheless, perhaps in certain areas of the brain where important decisions are made, they would be arranged in a square network.

4. CLASSIFICATIONS

The classification of James is that in his *Treatise on Psychology* (1912). Gregory has presented his classification in several articles, including 1991. For the verbal-satiation effect and the experiments of B. F. Skinner, see Warren (1982).

"We are here. . . ." According to the memoirs of Jean-Sylvain Bailly, the sentence really uttered was less ambiguous: "Go and say to those who send you that there's nothing the force of bayonets can do against the nation's will!" (See Arago, 1853.) The irrefutable paradox "it would be necessary to kill me . . ." was provided to me by the late Anne Guyon.

The quotation from Montaigne is in *Les Essais, Apologie de Raimond Sebond* (Editions Garnier-Flammarion, 1969). For Buffon, see Binet and Roger (1977). The development concerning the false illusion of the points is inspired by the thesis of Foucault (1910). The motif of Figure 4-5, bottom, belongs to a family of patterns, by Sakusi, displayed on the web site: http://psywww.human.metro- u.ac.jp/sakusi/gallery/gall_e.htm

5. LIMITS

In studying the effect of luminous persistence, Joseph Plateau was led to imagine, around 1830, the first devices for seeing sequences of images, giving an effect of continuous movement. Here too, one can consult his bibliography of 1878. An effect of instability in the manner of MacKay had been noted by Helmholtz for a figure with concentric circles (Figure 7–2). According to Michel Imbert (1987), the illusion of streaming has its origin in the processes of communication and cooperation between the brain cells responsible for detecting orientations.

For the discs of Fechner and Benham, I based my work on von Campenhausen and Schramme (1995). See also Viénot and Le Rohellec (1992). The illusion of Pulfrich's pendulum is generally attributed to poor stereoscopic estimation of the successive positions of the pendulum. I do not rule out that it may instead be a matter of the poor estimation of the components of speed, from which would result an error concerning the movement's direction.

6. CONTRASTS

The observations of Diquemare, Tycho Brahé, Gassendi, and La Hire were reported by Plateau (1878). The example of the black cat is taken from Arnheim (1989). For the Mach bands and the Craik-O'Brien effect, see Ratliff (1972). For a recent synthesis on Hermann's grid, see Spillmann (1994). Taya, Ehrenstein, and Cavonius (1995) have done the background history of the Munker-White effect. The analyses on the whiteness of the snow and the greenness of grass follow Koenderink and Richards (1992) and Pomerantz (1983). The wearing of yellow goggles for skiing was discussed by Kinney et al. (1983), and by many authors after them. The visibility of the letters under a transparent sheet of paper was the subject of a brief communication by Wilkins (1996).

When a series of guitar notes is rapidly played, each chord plucked continues to vibrate well after the next chord is activated and, consequently, the notes of the two chords are superimposed. Nevertheless, one hears a succession of detached notes. Craik and O'Brien constructed the visual illusion that bears their name by analogy with an older auditory illusion, that of Rawdon-Smith and Grindley (1935). When a sound of uniform intensity undergoes a sudden increase, followed by a gradual return to the initial intensity, the listener perceives a sudden increase in intensity, with no return to the initial intensity.

7. SEGREGATIONS, FUSIONS

O'Regan developed his criticism of the "fusion of images" in an article of 1992. It seems to me that there is indeed fusion; there remains the problem of determining at what stage of processing fusion occurs. The auditory segregations have been described in many articles and books; see notably Risset (1986) and Bregman (1990). I have presented the effect of symmetry with random patterns under the name "onirograms" in Ninio (1996a). We find other effects of symmetry in Wade (1982) and many examples of images within images, in the spirit of Figure 7–4, but much more elaborated in Wade (1990). On the perception of symmetry one can consult the book of Tyler (1996). Ouchi's illusion came into the scientific community when it was noticed by Lothar Spillmann and was chosen to appear on the cover of one of his books. The main properties of the illusion were established by the Italian researchers Bruno and Bressan (1994). I have constructed a number of original variants of them, and in particular have shown that areas with different orientations could migrate as a whole, and that two areas of an image could float simultaneously in different directions (Ninio, 1996b, Tourbe, 1996). The canvas of Reginald Neal is reproduced in Thurston and Carraher (1966).

Camouflage, whether for animals or the military, mainly exploits the effects of segregation and fusion. The distribution of color or shades of gray on a surface gives the illusion that the surface is fragmented, and the fragments are perceptually attached to the closest background areas. Hutchinson (in Gregory and Gombrich, 1973) calls attention to the fact that it is not necessary for animal camouflage to be absolutely effective: if it provides greater chances for survival, be it only 5 percent, that would be largely sufficient for it to become widespread in the species.

8. COMPLETIONS, CREATIONS

Ampère introduced the word "cognition" as the union of sensation with the sense of self (which he called "autopsy") in a letter of 1805 to Maine de Biran, in which he wrote: "This union of the autopsy with a sensation or intuition, just as with every other phenomenon, for autopsy unites with all of them, is what I call cognition; see if this word will not be serviceable for you, instead of perception." This letter is reproduced in the *La philosophie des deux Ampère* (1866). The extract on the Italian opera is found in it, reported by Ampère's son. In another letter to Maine de Biran, Ampère refutes the idea of the primacy of touch over vision, with the following argument: one can set a body in front of a mirror such that "when one brings the body closer to another body's image in the mirror, we seem to see two bodies that reciprocally penetrate each other and finally merge into one; which could not take place if we judged the relative positions of bodies according to what touch has taught us, for this sense could never have given us the idea of anything similar."

The quotation from Radau is taken from his *Acoustique* (1867). See Warren (1982) and Vicario (1982) for the experiments in auditory restoration. Illusory contours continue to engender large numbers of scientific articles every year. For a first approach, one can consult Petry and Meyer (1987) and Spillmann and Dresp (1995).

In the sonnet of colored vowels, Rimbaud also evoked the blue O and the green U. For the synesthesias, I took as my basis a series of article by Baron-Cohen et al., published in *Perception*, notably that of 1993.

The dream could have been taken up in this chapter, but as I have treated the forms taken by dream imagery in a rather detailed way in my *L'Empreinte des sens*, I refer the reader to that book for everything concerning it. Among related phenomena, there is the hypnagogic dream that occurs at the moment of dozing off. These images replace each other at stunning speed, probably at the natural rhythm of associations in the brain in the waking state. But in the hypnagogic dream, in general, we are conscious only of the final element evoked, and we are unable to reconstruct the intermediary steps leading to it. See the very acute analyses of Maury (1878). Let us also mention the "hypnopompic hallucinations," images from the first instant of awakening; see, for example, Shepard (1990) for some drawings representing his visions on awakening.

9. ADAPTATIONS

The history of the elevated subway is told in Karl Pribram (1969). The quotation from Goethe is taken from his *Treatise on Colors*, which contains numerous observations on the perception of color and experimental procedures, but scarcely furthers our understanding of the physics of light phenomena.

The experiment on the afterimage of a vowel is cited by Bregman (1990), who attributed it to Quentin Summerfield.

For the visual adaptive effects, see Harris (1980) and Rock and Harris (1967). As regards the presentation of Stratton's experiments, I have followed the account of Carr (1935). See also Kohler (1962) for his analogous experiments.

The first known description of adaptation to movement is due to Aristotle: if after gazing at moving objects—for example, a rapidly flowing river—one shifts one's sight to resting objects, the latter seem to move. And, from this observation, Aristotle outlined a theory of dreams: "These preserved impressions are the cause of dreams; they make themselves felt with greater intensity during sleep, when the mind and the senses are no longer active" (quoted by Plateau).

The history of the astigmatic was told to me by André Sentenac.

10. CONSTANCIES

The quotation from Fontenelle is one of the very first sentences of his *Entretiens sur la Pluralité des mondes* of 1686. The book by Maffei and Fiorentini (1995) is an accessible and richly illustrated initiation to visual perception. The writing by Buffon is taken from Binet and Roger

(1977). Constancy toward the top poses a problem when you want to write text on a wall: must you enlarge, and by how much should you enlarge the upper lines? See Baltrusaitis (1985).

Size constancy has posed problems for painters and draftsmen. Must the size of the moon be represented as it appears, or as it is measured? The "objective" representation, according to the measurements, gives a moon that is too small, because the brain is not enlarging the moon, represented on the picture, as it would enlarge a natural moon. Conversely, a moon put on the picture, with the diameter that the painter is tempted to give it, looks inordinately large. Another lunar paradox known to astronomers: when you search for the moon with the auxiliary lens of a refracting or reflecting telescope that has a good enlarging power, the moon looks smaller than it does to the naked eye!

Vergnaud's *Manuel de Perspective* (the second edition dates from 1826, the final one from 1881) contained reflections on perception that were quite interesting for its time. Notably, he stresses that the tendency observed in certain painters to use certain colors rather than others cannot be explained by defects in vision: "How to claim indeed that we can tell from a painter's work the way he perceives colors? Will he not, for example, use the color red, which he sees as brown, to represent the red object that he sees as brown? And, what's more, how will he tip us off to such an oddity, since he will necessarily use the same word to express a color that he sees differently? He will take an object he was taught was red, and although he sees it as brown, he will say there is red. . . ." See also, in the same spirit, the quotation from Rohault in Chapter 13. Unfortunately, Vergnaud did not make his manual evolve in its successive editions, whereas the psychology of vision progressed explosively.

On the history of the raising of the tuning-fork pitch, see Scotto di Carlo (1997).

11. Reference Points

Rohault maintained that interposition cues operated in the moon illusion. "But when it has just risen or is ready to set, we see between it and us several countrysides whose size we more or less know and thus we judge it as farther away, and because of this we see it as larger." But Le Cat (1742) said that the size illusion persists when one aims at the moon through a tube, without any interposition in the field of the tube. The illusion of the "crow hunter" and of the houses in the Landes forest are described in the precise and relevant book by Armand de Gramont (1939). The observations of Harvey Carr are recorded in three articles, including Carr (1909). The illusion of the terrain in Banyuls was the subject of a press report by J. Lopez (1991).

The observations of Ernst Mach, one of the most eminent thinkers on the relativity of movement, are recorded in the *Analysis of Sensations*, but he was duped by the "illusion of the pigeon's head." Bourdon's book (1902) on the perception of space is one of the best French contributions of all time to the domain of visual perception. The author appears in it with the modest title of professor of philosophy at the University of Rennes. In France, at the beginning of the twentieth century, there was thus already a separation between the institutional hierarchy and that of talent.

As regards clouds seen from an airplane, Arnheim's observation is taken from his book of 1989. Shipley (1976) described a particularly complex situation: one glorious day, flying at an altitude of 30,000 feet, he saw much farther down two sets of overlapping clouds, both scattered, not thick and of about the same appearance, with irregular movements. Gradually, the two sets separated; one set accompanied the plane as if it were attached to it. The other set, showing through the first one, seemed to move in the direction opposite that of the plane. Trying to catch a glimpse of the ocean, the author ended up making out its surface, still lower down, and saw it move as well, as if pulled by the plane. At that moment, conscious of the impossibility of the situation, he thought of another interpretation: the layer that moved in the direction opposite to the plane continued in that direction but became the nearest cloud layer while the layer that accompanied the plane became the cloud layer below, close to the ocean, and both of them together soon started to drift slightly in the direction opposite the plane. On the topics of the appearance of telephone poles when riding in a car, and the snow that flies at the windshield, see Farné and Sebellico (1985).

12. ARBITRATIONS

On the fact that vision is not educated by touch, see in the notes to Chapter 8 Ampère's argument concerning the interpenetration of two objects. The conflicts between touch and vision are discussed well in Rock and Harris (1967). Warren (1982) has treated audition in relation to the other senses, and notably the experiment of McGurk and MacDonald. Another example of the influence of vision on other senses is that of the discomfort that affects many people on car trips, especially on turns—a sensation similar to seasickness. This sensation is primarily felt by passengers in the back seat who are reading or not looking attentively at the road. The interpretation is that seeing the road makes it possible to anticipate the bodily sensation of centrifugal force or acceleration. Seasickness is felt when these sensations take you by surprise.

My interest in geometric illusions is an old one: I published my theoretical analysis, in which I show that most of the metric illusions boil down to three principles, in an article of 1979. Later, I developed an experimental approach and studied with Kevin O'Regan new variants of the Zöllner and Poggendorff illusions (Ninio and O'Regan, 1996, 1998). That experimental psychology is mystified to such a degree by geometric illusions primarily indicates the absence of geometric intuition prevailing in this field, a situation that can only grow worse, considering how the computerized "cookbook" approach to data is supplanting mathematical concepts. The chapter on illusions published by Bourdon in his book (1902) is among the most pertinent I know on the subject. The Japanese contributions to the domain of geometric illusions and, in particular, the Morinaga paradox, are presented in the review article of Oyama (1960).

13. ILLUSIONS AND CULTURE

The theme of the different interpretations of images in different cultures has been discussed by Deregowski in numerous books. One can, in particular, consult his 1989 article, which is introductory to a debate, followed by numerous critical commentaries by specialists. One element often forgotten in intercultural comparisons is that, because of the inequalities in medical care, the members of certain communities often have important uncorrected defects in vision. The distortions of perspective noticeable on photos, and particularly the example of the sphere, were discussed by Pirenne (1970). Segall is the author of two books, but his article of 1963 gives what is essential, it seems to me.

Rohault's passage on his color deficiency is remarkable: he anticipates the discovery of color blindness. Malebranche predicted, as a personal speculation, the existence of persons who could have the deficiency . . . precisely described by Rohault.

One would tend to think that writing, and notably the fact of writing from right to left in certain cultures, from left to right in others, would strongly influence visual perception. The only convincing piece of supporting information concerns the assessment of the middle of segments which the French locate too far to the left, and the Israelis (who read Hebrew from right to left) place too far to the right (Chokron and Imbert, 1993).

Certain geometric illusions have an optional character. Everything happens as if an area of the brain correctly assessed the relations while another one brought a distorted representation to consciousness. This is manifest in stereoscopic vision in which an illusion present in a two-dimensional figure can disappear in three-dimensional vision (Ninio, 1994). This is also observed in tests of grasping, in which we correctly separate our fingers to grasp an object, even if the object is perceived as having the wrong size (Goodale and Milner, 1992).

14. ILLUSIONISM

The experiments of "spiritism" (pseudo-dialogues with the dead) appeared, according to Girden (1978), in 1848 in a farm in upstate New York. The mediums were two girls aged six-and-a-half and eight. Forty years later, the elder one, Margaret Fox, acknowledged that she had given her responses by tapping with her knuckles and, later, with the joint of her big toe. Girden also discusses Uri Geller and miniature transmitter-receivers. I have witnessed the trick of substituting actors for puppets during a dance performance of Philippe Genty. The trick of the pigeon replacing the balloon was performed by the magician David Jonathan Bass during a televised 1997 "Golden Mandrake" contest, which he won.

15. ARTIFICE AND CONVENTION

I could have called this chapter "Art and Illusion," but because the theme was developed in the latest edition of *L'Empreinte des sens*, I have chosen here to take it up in a more down-to-earth way, discussing movie techniques, for movies affect a larger public today than altar pieces and icons.

The worldwide infatuation with sports information stems from the fact that, even if the events are rigged, the report of them is accurate: if it is said that a particular team has won the game, that a particular player has scored the first goal, these are universal truths that can be shared by everyone on the planet. Sports have thus become one of the very rare domains—along with obituaries and daily weather and stock reports—where the information is, on the whole, still truthful. On perceiving relief with one eye, see Claparède (1904), Ames (1925), and Koenderink, Van Doorn, and Kappers (1994).

For the history of the panorama and the diorama, see Bapst (1891) and Kemp (1990), and for the synopter, see Koenderink, Van Doorn, and Kappers (1994). Among the techniques contributing to the effectiveness of the panorama, Bapst cites the "parajour" (overhead screen): "A parajour, located above the viewer, hides what is above his head, prevents him from seeing the upper end of the painting and the circular opening through which daylight enters; glare is thus lessened, and the viewer's shadow is not cast on the canvas." Moreover, "to bring the viewer from outside up to the platform, he is led through dark corridors; on the way, he loses the idea of light and when he arrives at his assigned place he goes without transition from darkness to the sight of the circular painting exposed to the brightest light: then, all points of the panorama are presented at the same time and there results a kind of confusion; but soon, as the eye becomes accustomed to the brightness, the painting subtly produces its effect, and the more the viewer stares at it, the more persuaded he is that he is in the presence of reality."

On the theme of visual perception and the cinema, I have used Hochberg and Brooks (1978) a bit and especially my own memories.

16. ILLUSIONS OF THE MEMORY & THE MIND

An experiment of von Békésy: vibrators all at the same frequency put pressure on the end of the index and the middle finger. The vibrators are applied in such a way that the sensation has the same intensity on the two fingers considered separately. If the impulses are separated by three or four milliseconds, the person feels separate sensations in the two fingers. If the interval between the impulses is reduced to one millisecond, the two vibrations merge and the sensation is localized in the finger receiving the impulses with this millisecond advance. If the interval between the impulses is reduced even further, then reversed, the person will have, at the moment of the reversal, a sensation of pressure that suddenly jumps from the end of one finger to the end of the other. After two or three weeks of tasks of this sort, however, the sensation of a jump is replaced by that of a progressive displacement from one finger to the other. Finally, when the impulses arrive at exactly the same time, the subject feels the pressure midway between the two fingers. For the section on the phantom limbs, I used Melzack (1992) and a lecture given by Ramachandran in Helsinki (1997). The experiments of the weighing of numbers are described in Dehaene (1997). There exist numerous illusions of memory, including the "déjà vu" illusion and that of the "jamais vu" [never seen], the incorporation of false recollections in memory, the illusion of "I knew it all along" when we are given the answer to a question on which we had drawn a blank, and so forth. See, for example, the review article by Roediger (1996). The experiment described in the text, of incorporating a word in a list, is reported here following the article by Don Read (1996).

BIBLIOGRAPHY

Ames, A. 1925. The illusion of depth from single pictures. *J. Optical Soc. America* 10: 137–48.

Ampère, A.M., and J.J. Ampère. 1866. *Philosophie des deux Ampère.* Paris: Didier.

Arago, D. F. J. 1853. Biographie de Jean-Sylvain Bailly. In *L'Annuaire pour l'an 1853*, 343–630. Paris: Bachelier.

Arnheim, R. 1989. *Parables of Sunlight: Observations on Psychology, the Arts, and the Rest.* Berkeley: University of California Press.

Baltrusaitis, J. 1977. *Anamorphic Art.* Translated by W. Strachan. New York: Harry N. Abrams.

Bapst, G. 1891. Essai sur l'histoire des panoramas et des dioramas. In *Prehistory of Photography: Five Texts.* Edited by Robert Sobieszek. New York: Arno Press.

Baron-Cohen, S., J. Harrison, L. Goldstein, and M. Wyke. 1993. Colored speech perception: Is synesthesia what happens when modularity breaks down? *Perception* 22: 419–26.

Békésy, G. von. 1967. *Sensory Inhibition.* Princeton: Princeton University Press.

Bergen, J. R. 1988. Hermann's grid: new and improved. *Investigative Ophthalmology and Visual Science* S26: 280.

Binet, J.L., and J. Roger. 1977. *Un autre Buffon.* Paris: Hermann.

Biot, J.B. 1820. Note adressée à M. Biot par feu M. Jurine, de Genève, sur un phénomène de mirage latéral. *Bulletin de la Société de philomatique* 10: 28–31.

Boring, E. G. 1942. *Sensation and perception in the history of experimental psychology.* New York: Irvington, 1977.

Bouhot, G. 1994. Surprises . . . à la lecture d'images. *Phot Argus* 198: 51–56.

Bourdon, B. 1902. *La Perception visuelle de l'espace.* Paris: Schleicher Frères.

Bregman, A. S. 1990. *Auditory scene analysis: The perceptual organization of sound.* Cambridge: MIT Press.

Bruno, N., and P. Pressan. 1994. Paradoxical motion in stationary patterns. *Perception* 23 (ECVP supplement 94): 28.

Burton, H. E. 1945. The optics of Euclid. *J. Optical Soc. America* 35: 357–72.

Campenhausen, C. von, and J. Schramme. 1995. 100 years of Benham's top in color science. *Perception* 24: 695–717.

Carlson, C. R., C. H. Anderson, and J. H. Moeller. 1980. Visual illusions without low spatial frequencies. *Investigative Ophthalmology and Visual Science* 19: S 165.

Carr, H. A. Visual illusions of depth. *Psychol. Rev.* 16: 219–56. See also 13: 258–75 and 15: 139–49.

——. 1935. *An introduction to space perception.* New York: Longmans and Green.

Chokron, S., and M. Imbert. 1993. Influence of reading habits on line bisection. *Cognitive Brain Res.* 1: 219–22.

Claparède, E. 1904. Stéréoscopie monoculaire paradoxale. *Annales d'oculistique* 132: 465–66.

Coren, S., and J. S. Girgus. 1978. *Seeing is deceiving: The psychology of visual illusions.* Mahwah, N. J.: Lawrence Erlbaum.

Coren, S., and C. Porac. 1983. The creation and reversal of the Möller-Lyer illusion through attentional manipulation. *Perception* 12: 49–54.

Da Pos, O., and E. Zambianchi. 1996. *Visual illusions and effects-a collection.* Milan: Guerini Studio.

Day, R. H., and E. J. Stecher. 1991. Sine of an illusion. *Perception* 20: 49–55.

Dehaene, S. 1997. *The Number Sense: How the Mind Creates Mathematics.* New York: Oxford University Press.

Deregowski, J. B. 1989. Real space and represented space: Cross-cultural perspectives. *Behavioral and Brain Science* 12: 51–119.

Deutsch, D. 1978. The psychology of music. In Carterette and Friedman, eds., *Handbook of perception* vol. 10: *Perceptual ecology,* 191–224.

Don Read, J. 1996. From a passing thought to a false memory in 2 minutes: Confusing real and illusory events. *Psychonomic Bulletin & Review* 3: 105–11.

Enright, J. T. 1970. Distortions of apparent velocity: A new optical illusion. *Science* 168: 464–67.

Farné, A., and A. Sebellico. 1985. Illusory motions induced by rapid displacements of the observer. *Perception* 14: 393–402.

Foucault, M. 1910. *L'Illusion paradoxale et le seuil de Weber.* Montpellier: Coulet et fils.

Fraser, J. 1908. A new illusion of visual direction. *British J. Psychol.* 2: 307–20.

Girden, E. 1978. Parapsychology. In Carterette and Friedman, eds. *Handbook of Perception* vol. 10: *Perceptual ecology:* 385–412. New York: Academic Press.

Goethe, J. W. von. 1810. *Theory of colors.* Translated by Charles Lock Eastlake. Cambridge: MIT Press, 1970.

Goodale, M. A., and A. D. Milner. 1992. Separate visual pathways for perception and action. *Trends in Neurosciences* 15: 20–25.

Gramont, A. de. 1939. *Problèmes de la vision.* Paris: Flammarion.

Gregory, R. L. 1963. Distortion of visual space as inappropriate constancy scaling. *Nature* 199: 678–80.

——. 1991. Putting illusions in their place. *Perception* 20: 1–4.

Gregory, R. L., and E. H. Gombrich, eds. 1973. *Illusions in nature and art.* London: Gerald Duckworth.

Hammersley, R. 1983. Things are deeper than they are wide: A strange error of distance estimation. *Perception* 12: 589–91.

Harris, C. S., ed. 1980. *Visual coding and adaptability.* Mahwah, N.J.: Lawrence Erlbaum.

Harris, J. P., and R. L. Gregory. 1973. Fusion and rivalry of illusory contours. *Perception* 235–47.

Helmholtz, H. von. 1856–1866. *Helmholtz's treatise on physiological optics* New York: Dover, 1967.

Hochberg, V., and V. Brooks. 1978. The perception of motion pictures. In Carterette and Friedman, eds., *Handbook of Perception,* vol. 10: *Perceptual ecology:* 259–304.

Howard, I. P. 1966. Alhazen's neglected discoveries of visual phenomena. *Perception* 25: 1203–17.

Howard, I. P., and N. J. Wade. 1996. Ptolemy's contributions to the geometry of binocular vision. *Perception* 25: 1189–201.

Idesawa, M., and Q. Zhang. 1997. Occlusion cues and sustaining cues in 3–D illusory object perception with binocular viewing. *SPIE Proceedings* 3077: 770–81.

Imbert, M. 1987. Vous n'en croyez pas vos yeux! *Le Courrier du CNRS,* supplement to numbers 66–67–68: January–June, p. 2 of cover.

James, W. 1912. *The principles of psychology.* Cambridge: Harvard University Press, 1983.

Kanizsa, G. 1980. *Grammatica del vedere.* Bologna: Il Mulino

Kemp, M. 1990. *The science of art: Optical themes in Western art from Brunelleschi to Seurat.* New Haven: Yale University Press.

Kepler, J. 1604. *Gesammelte Werke.* Edited by Walther von Dyck and Max Caspar. 18 vols. (in progress). Munich, 1937–. Vol. 2 (1939): *Ad Vitellionem paralipomena, quibus astronomiae pars optica traditur.* Edited by Franz Hammer. Vols. 13–18 (1945–59): Briefe 1590–1630.

Kinney, J. A., S. M. Luria, C. S. Schlichting, and D. F. Neri. 1983. The perception of depth contours with yellow goggles. *Perception* 12: 363–66.

Koenderink, J., and W. A. Richards. 1992. Why is the snow so bright? *J. Optical Soc. America.* A 9: 642–48.

Koenderink, J. J., A. J. Van Doorn, and A. M. L. Kappers. 1994. On so-called paradoxical monocular stereoscopy. *Perception* 23: 583–94.

Kohler, I. May 1962. Experiments with goggles. *Scientific American* 206: 63–72.

La Hire, P. de. 1694. *Mémoires de mathématiques et de physique. Un traité des différents accidents de la vue.*

Landroy, H. D. 1922. *Dernières pensées d'un décapité.* Louvain: Femina Press.

Lanners, E. 1977. *Illusions.* Translated by Heinz Norden. London: Thames and Hudson.

Le Cat, C.-N. 1742. *Traité des sens.* Amsterdam: J. Wetstein.

Leipp, E. 1977. *La machine à écouter. Essai de psycho-acoustique.* Paris: Masson.

Lejeune, A. ed., 1956. *L'Optique de Claude Ptolémée dans la version Latine d'après l'Arabe de l'Émir Eugène de Sicile.* Louvain: Université de Louvain.

Leviant, I. 1996. Does "brain-power" make *Enigma* spin? *Proceedings of the Royal Soc.* B 263: 997–1001.

Lindberg, D. C. 1976. *Theories of vision from al-Kindi to Kepler.* Chicago: University of Chicago Press.

Logvinenko, A. D. 1999. Lightness induction revisited. *Perception* 28: 803–16.

Lopez, B. H. 1989. *Arctic dreams: Imagination and desire in a northern landscape.* New York: Bantam Books.

Lopez, J. 1991. La montée qui descend. *Science & Vie Junior* 32: 12–16.

Lukiesh, M. 1922. *Visual illusions: Their causes, characteristics, and applications.* New York: Dover.

Mach, E. 1900. *The analysis of sensations, and the relation of the physical to the psychical.* Translated by C. M. Williams. New York: Dover, 1959.

MacKay, D. M. 1957. Moving visual images produced by regular stationary patterns. *Science* 149: 849–50.

Maffei, L., and A. Fiorentini. 1995. *Arte e cervello.* Bologna: Zanichelli.

Malebranche, N. 1712. The search after truth: [with] elucidations. Translated by Thomas Lennon and Paul J. Olscamp. New York: Cambridge University Press, 1997.

Mariotte, E. 1684. *Lettres écrites sur le sujet d'une nouvelle découverte touchant la vue, faite par M. Mariotte.*

Maury, A. 1878. *Le Sommeil et les rêves.* 4th ed. Paris: Didier.

McBeath, M. K. 1990. The rising fastball: Baseball's impossible pitch. *Perception* 19: 545–52.

Mehler, J., and E. Dupoux. 1994. *What infants know: the new cognitive science of early development.* Translated by Patsy Southgate. Cambridge, Mass.: Blackwell.

Melzack, R. April 1992. Phantom limbs. *Scientific American* 266: 90–96.

Metzger, W. 1975. *Gesetze des Sehens.* Frankfurt: Waldemar Kramer.

Morinaga, S., and H. Ikeda. 1965. Paradox in displacement in geometrical illusion and the problem of dimensions: A contribution to the study of space perception. *Japanese J. Psychol.* 36: 231–38.

Münsterberg, H. 1894. *Pseudoptics.* New York: Milton Bradley.

Ninio, J. 1979. An algorithm that generates a large number of geometric visual illusions. *J. Theoret. Biol.* 79: 167–201.

——. 1994. La vision stéréoscopique, sens méconnu. *Pour la Science* 197: 28–33.

——. 1996a. Onirogrammes. *Pour la Science* 221:108–09.

——. 1996b. Flottements. *Pour la Science* 223: 95–96.

Ninio, J., and J. K. O'Regan. 1996. The half-Zöllner illusion. *Perception* 25: 77–94.

Ninio, J., and J. K. O'Regan. 1999. Characterization of the misalignment and misangulation components in the Poggendorff and corner-Poggendorff illusions. *Perception* 28: 949–964.

Ninio, J., and K. A. Stevens. Variations on the Hermann grid: A curious extinction illusion. *Perception* 29: 1209–1217.

O'Regan, K. 1992. Solving the "real" mysteries of visual perception: The world as an outside memory. *Canadian J. Psychol.* 46/3: 461–88.

Ouchi, J. 1977. *Japanese optical and geometrical art.* New York: Dover.

Oyama, T. 1960. Japanese studies on the so-called geometrical-optical illusions. *Psychologia* 3: 7–20.

Ozanam, J. 1692. *Récréations mathématiques et physiques.* New edition, Paris: chez Charles-Antoine Jombert, 1750.

Papert, S. 1961. Centrally produced geometrical illusions. *Nature* 191:733.

Petry, S., and G. E. Meyer, eds. 1987. The perception of illusory contours. New York: Springer.

Pirenne, M. H. 1970. *Optics, painting and photography.* New York: Cambridge University Press.

Plateau, J. 1878. Bibliographie analytique des principaux phénomènes subjectifs de la vision, depuis les temps anciens jusqu'à la fin du XVIIIe siècle, suivie d'une bibliographie simple pour la partie écoulée du siècle actuel. *Mémoires de l'Académie royale des sciences, des lettres et des beaux-arts de Belgique, vol. 52: Section 1: Persistance des impressions sur la rétine. Section 2: Couleurs accidentelles ordinaires de succession. Section 3: Images qui succèdent à la contemplation d'objets d'un grand éclat ou même d'objets blancs bien éclairés. Section 4: Irradiation. Section 5: Phénomènes ordinaires de contraste. Section 6: Ombres colorées.*

Polyak, S. 1941. *The retina.* Chicago: University of Chicago Press.

Pomerantz, J. R. 1983. The grass is always greener: An ecological analysis of an old aphorism. *Perception* 12: 501–02.

Pribram, K. H. January 1969. The neurophysiology of remembering. *Scientific American* 228: 73–86.

Proffitt, D. R., M. Bhalla, R. Gossweiller, and J. Midgett. Perceiving geographical slant. *Psychonomic bulletin & Review* 2: 409–28.

Radau, R. 1867. *L'Acousticque, ou les phénomènes du son.* Paris: Hachette.

Raj Maharnipur, L. H. 1967. *Conceptual foundations of ventriloquism.* Kalamazoo: Western Michigan University Press.

Ramachandran, V. S. May 1992. Blind spots. *Scientific American* 266: 44–49.

——. 1997. Synesthesia and external "projection" of kinesthetic sensations in phantom limb patients and normal individuals. *Perception* 26 (supplement): 69.

Ratliff, F. June 1972. Contour and contrast. *Scientific American* 226: 90–101.

Rawdon-Smith, A. F., and G. C. Grindley. 1935. An illusion in the perception of loudness. *Brit. J. Psychol.* 26: 191.

Risset, J.-C. 1986. Son musical et perception auditive. *Pour la Science* 109: 32–43.

Robinson, J. O. 1972. *The psychology of visual illusions.* London: Hutchinson University Library.

Rock, I., and C. S. Harris. May 1967. Vision and touch. *Scientific American:* 96–104.

Roediger, H. L. 1996. Memory illusions. *J. Memory and Language* 35: 76–100.

Rohault, J. 1681. *Rohault's system of natural philosophy.* Translated by James Clark. New York: Garland, 1987.

Rudaux, L. 1937. *Sur les autres mondes.* Paris: Larousse.

Sabra, I. 1989. *The optics of Ibn al-Haytham.* London: Warburg Institute.

Savigny, G. B. de. 1905. *Les amusements de la science.* Paris: Librarie des publications populaires.

Schrauf, M., B. Lingelback, and E. R. Wist. 1997. The scintillating grid illusion. *Vision Research* 37: 1033–38.

Scotto di Carlo, N. 1997. Les divas donnent le *la. Pour la Science* 323: 24.

Segall, M. H., D. T. Campbell, and M. J. Herskovits. 1963. Cultural differences in the perception of geometric illusions. *Science* 139: 769–71.

Shepard, R. N. 1990. *Mind sights.* New York: Freeman & Co.

Shipley, T. 1976. Flying clouds: An illusion of visual capture and distance reversal. *Vision Res.* 16:1522–24.

Spillman, L. 1994. The Hermann grid illusion: A tool for studying perceptive field organization. *Perception* 23: 691–708.

Spillman, L., and B. Dresp. 1995. Phenomena of illusory form: Can we bridge the gap between levels of explanation? *Perception* 24: 1333–64.

Stratton, G. M. 1887. Vision without inversion of the retinal image. *Psychological Reviews* 4: 341–60 and 463–81. See also: 1886. 3: 611–17.

———. 1899. The spatial harmony of touch and light. *Mind* 8: 492–505.

Taya, R., W. H. Ehrenstein, and R. C. Cavonius. 1995. Varying the strength of the Munker-White effect by stereoscopic viewing. *Perception* 24: 685–94.

Thurston, B. J., and G. Carraher. 1966. *Optical illusions and the visual arts.* New York: Van Nostrand Reinhold.

Tolansky, S. 1964. *Optical illlusions.* London: Pergamon.

Topper, D. R. 1984. The Poggendorff illusion in *Descent from the Cross* by Rubens. *Perception* 13: 655–58.

Tourbe, C. 1996. L'illusion du mouvement. Illustrations by J. Ninio. *Science et Avenir* 594: 74–76.

Trouessart, J. 1854. *Recherches sur quelques phénomènes de la vision, précédées d'un essai historique et critique des théories de la vision, depuis l'origine de la science jusq'à nos jours.* Brest: Imprimerie Édouard Anner.

Tse, P. U. 1998. Illusory volumes from conformation. *Perception* 27: 977–91.

Tyler, C. W., ed. 1996. *Human symmetry perception and its computational analysis.* Igness. Zeist, The Netherlands: VSP.

Varin, D. 1971. Fenomeni di contrasto e diffusione cromatica nell'organizzazione spaziale del campo percettivo. *Rivista di Psicologia* 65: 101–28.

Vergnaud, A. D. 1826. *Manuel de perspective du dessinateur et du peintre.* Paris: Roret.

Vicario, G. B. 1972. Phenomenal rarefaction and visual acuity under "illusory" conditions. *Perception* 1: 475–82.

Viénot, F., and J. Le Rohellec. 1992. Reversal in the sequence of the Benham colors with a change in the wavelength of illumination. *Vision Res.* 12: 2369–74.

Wade, N. 1982. *The art and science of visual illusions.* London: Routledge and Kegan Paul.

——. 1990. *Visual allusions: Pictures of perception.* Mahwah, N.J.: Lawrence Erlbaum.

——. 1996. Descriptions of visual phenomena from Aristotle to Wheatstone. *Perception* 25: 1137–75.

Warren, R. M. 1982. *Auditory perception: A new synthesis.* New York: Pergamon Press.

Wessel, D. L., and J.-C. Risset. 1979. Les illusions auditives. In *Universalia* 1979: 161–71. Paris: Encyclopaedia Universalis.

Wilkins, A. 1996. Helping reading with color. *Perception* 25 (supplement): 74.

Zavagno, D. 1999. Some new luminance-gradient effects. *Perception* 28: 835–38.

Zwicker, E. 1964. Negative afterimage in hearing. *J. Acoust. Soc. Amer.* 36: 2413–15.